Scale, Heterogeneity, and the Structure
and Diversity of Ecological Communities

MONOGRAPHS IN POPULATION BIOLOGY
EDITED BY SIMON A. LEVIN AND HENRY S. HORN

Complete series list follows Index

Scale, Heterogeneity, and the Structure and Diversity of Ecological Communities

MARK E. RITCHIE

PRINCETON UNIVERSITY PRESS
Princeton and Oxford

Published by Princeton University Press,
41 William Street, Princeton, New Jersey 08540
In the United Kingdom: Princeton University Press,
6 Oxford Street, Woodstock, Oxfordshire OX20 1TW

Library of Congress Cataloging-in-Publication Data

Ritchie, Mark E., 1960–
  Scale, heterogeneity, and the structure and diversity of ecological communities /
Mark E. Ritchie.
   p. cm. — (Monographs in population biology)
  Includes bibliographical references and index.
  ISBN 978-0-691-09069-6 (hardcover : alk. paper)—ISBN 978-0-691-09070-2
(pbk. : alk. paper)   1. Ecological heterogeneity.   2. Biodiversity.   3. Biotic
communities.   4. Animal population density.   I. Title.
QH541.15.E24R58 2009
577—dc22   2009029065

British Library Cataloging-in-Publication Data is available

Printed on acid-free paper. ∞

press.princeton.edu

Printed in the United States of America

10   9   8   7   6   5   4   3   2   1

Contents

Acknowledgments

This work originated out of visiting fellowships from the Netherlands Scientific Research Organization (NWO) and the Santa Fe Institute and was greatly helped to its conclusion by a visiting fellowship from the U.S. National Science Foundation-supported National Center for Ecological Analysis and Synthesis in Santa Barbara. Collection of original field data was supported by several grants from the National Science Foundation and by funds from Syracuse University. Many people stimulated ideas in the book, but I especially thank Han Olff as a collaborator on several recent papers with related topics, and James H. Brown, Jessica Green, Matt Leibold, Brian Maurer, Bruce T. Milne, and the attendants at several Santa Fe Institute workshops on biological complexity for comments. I thank V. Brent Davis, Will Pitt, John Whiteman, Lisa Krause, Melanie Harsch, Sean Kyle, Kerry Griffis-Kyle, Zhongan Chen, and numerous field interns at Cedar Creek Natural History Area of the University of Minnesota for their help in collecting field data. I thank Sumanta Bagchi for a great help in editing the final manuscript. Finally, I thank my wife Estelle and three children Connor, Noah, and Shelby for their patience with my absent-mindedness while thinking about this book.

CHAPTER ONE

Community Ecology Lives

Understanding what controls the structure and diversity of ecological communities has invoked the intellectual firepower of ecologists since at least the time of Charles Darwin (1859, p. 125).

In the case of every species, many different checks, acting at different periods of life, and during different seasons or years, probably come into play; some one check or some few being the most potent, but all concurring in determining the average number or even the existence of the species. . . . When we look at the plants and bushes clothing an entangled bank, we are attempted to attribute their proportional numbers and kinds to what we call chance. But how false a view is this!

Few conceptual undertakings in science, much less ecology, have such an ambitious goal that applies over such a wide range of scales in space (May 1988; Wilson 1992; Lawton 1999). Community ecology is difficult because, as Darwin recognized, many factors affect the existence and abundance of organisms. Ecologists now have explored in detail how resource availability, disturbance, dispersal, predation, disease, mutualism, evolutionary history, scale of observation, and variability of physical conditions and resources over space and time (heterogeneity) affect community structure. These different factors influence community structure at different scales in space and time. Simberloff (2004) has suggested there are few general principles; the goal for now should be the accumulation of case studies until we can understand the context of different outcomes of community organization. Even more pessimistically, Lawton (1999) suggested that we "move on" and essentially give up on a general theory, given that communities seem to have highly contingent, unpredictable patterns.

Different scientific approaches to understanding community structure emerged early in the development of ecology as a science (Kingsland 1988). Ecologists use a deductive approach to predict community
structure from particular mechanisms, such as competition for resources or dispersal (Tilman 1982, 1994; Belovsky 1986; Hubbell
2001; Chase and Leibold 2003). Ecologists adopting this approach
often develop analytical, mathematically derived hypotheses of interactions among species. These predictions are most often compared
with the results of laboratory and field experiments. The other approach
is more inductive, in that ecologists explore patterns in community
structure and compare them with broad, often non-mathematical, hypotheses that do not specify the mechanisms generating such patterns.
Each approach has provided major advances and insights (Brown
1981, 1995; Huston 1994; Ricklefs and Schluter 1994; Rosenzweig
1995), but they remain largely unreconciled. The difference in these
two approaches has separated ideas and data whose synthesis might
lead to a more thorough understanding of community structure. To
move forward, both deductive and inductive approaches must be
synthesized.

   In this book, I propose a new framework for predicting the structure
and diversity of ecological communities that might help synthesize
previous theory and data. This framework emerges out of incorporating
two critical elements of the inductive approaches, scale and heterogeneity, into the analytical mathematical formalism of the more deductive
approaches. The new framework makes novel predictions of diversity
that depend explicitly on the spatial and temporal scale of the observer,
the inherent heterogeneity of the environment in space, and the scale of
response to the environment by different species in space. It is a formal
extension of the original ideas of environmental and perceptual "grain"
introduced by Levins (1962, 1968) and MacArthur (1972).

   The emphasis on scale and heterogeneity requires a tool that can
simply describe the complex physical structure of nature: fractal geometry. Fractal geometry assumes that distributions of physical material and conditions and/or biological organisms in the environment are
statistically similar across a range of meaningful spatial scales (Mandelbrot 1982; Milne 1992). In that sense, it is a "neutral" model of
heterogeneity. It describes very complex-looking distributions with
simple mathematical scaling laws. By inserting these scaling laws

directly into classical population dynamics models, new models emerge that incorporate the scale-dependent description of spatial heterogeneity so critical to the interpretation of macroecological patterns (Brown 1995). Just as importantly, these models can only be interpreted in terms of the spatial scales of the species involved. By assuming that organism body size provides a first approximation to these scales (Peters 1983; Calder 1984; Charnov 1994; West et al. 1997; Brown et al. 2004), the models explicitly link the conditions for coexistence to species body size (Morse et al. 1985) and thus to species composition, abundance, and diversity. The utility depends on including both scale and a metric of spatial pattern in traditional models of resource dynamics and consumption. The approach would apply even to the many environments in which spatial patterns appear to vary across scales (Allen and Holling 2002), as the scaling laws inherent in the fractal geometric description of heterogeneity can be adapted to include exponents that are themselves functions of scale. For the purposes of this book, I devote my attention to how a simple assumption of fractal geometry, as a first approximation, can elucidate how species select and partition packages of the same limiting resource in order to coexist.

WHY ARE THERE SO MANY SPECIES?

To understand how such a framework might provide progress and synthesis, I return to 1959, when G. Evelyn Hutchinson posed the question, "Why are there so many species?" This deceptively simple question was novel then because the results of the previous century of natural history, ecological theory and experimentation led to a conundrum. Theory (Lotka 1925; Volterra 1926) and laboratory experiments (Gause 1934; Park 1948) suggested that coexistence occurred only under specific conditions. The interpretation of these results led to the "principle of competitive exclusion": no two species that are identical in their use of resources can coexist, and coexistence therefore occurs only under special conditions (Hardin 1960). However, this "principle" made little sense to natural historians, who confronted it with the observation of myriad coexisting species, many of which seemed to use similar resources (Elton 1958; Hutchinson 1957).

Hutchinson's question challenged the "principle of competitive exclusion," and thus challenged theoretical and experimental ecologists to determine how different species must be to coexist, and how these differences determine the large, but not infinite, number of species we observe in nature. Ecologists addressed this challenge in two major ways. First, population ecologists began the search for potential mechanisms that could allow multiple species to coexist on relatively few (or a single) limiting resources. Differences in diet overlap were interpreted as leading to differences in per capita competitive effects. When placed in classical Lotka-Volterra models, these per capita effects among all possible species pairs ultimately predicted a "community matrix" of interaction coefficients among multiple species (MacArthur 1970; Strobeck 1972; May 1976). Such differences in diet overlap among species were thought to emerge from the evolution of different "optimal" foraging strategies (MacArthur and Pianka 1966; Emlen 1966; Schoener 1971) in which individuals selected diet items to maximize fitness. Species would have different "optimal" diets, which would not completely overlap and therefore would promote their coexistence. Other work showed how pairs of species might coexist only when they differed in the size or type of resources used (MacArthur 1972; Schoener 1974; Tilman 1982). In contrast, MacArthur and Wilson (1967) suggested that species diversity in islands or fragmented habitats were controlled by colonization and extinction. All these approaches focused on how community structure emerged from the dynamics and evolution of populations near equilibrium population sizes, as dictated by competition with other species. Natural selection, when combined with competition, could yield differences among species in their morphological, physiological, and behavioral traits. These trait differences would allow them to use different *niches*, or combinations of physical locations, conditions, resources, and interactions with competitors and predators, thus promoting coexistence (Grant 1986; Thompson 1994; Leibold 1995; Chase and Leibold 2003). A comprehensive theory of the niche and the role of niches in predicting community structure dominated ecological thought by the late 1970s and provided a structure for interpreting a tremendous volume of natural history information.

Despite this progress, the population dynamic approach to community ecology could only vaguely predict the wealth of patterns in species abundance and diversity, particularly at larger spatial scales. This void was filled by community ecologists, who employed a more inductive approach. For example, Preston (1962) and MacArthur (1965) found predictable patterns in the abundance and diversity of different-sized species in communities. These results inspired other ecologists to search for patterns in other taxa and environments. Numerous studies yielded relationships between species diversity and the area of continents and islands, and relationships between diversity and productivity (Whittaker 1975), disturbance (Connell 1978), latitude (MacArthur 1965), and various measures of resource and habitat heterogeneity (e.g., MacArthur 1965). These patterns largely were explained by a myriad of verbal hypotheses or statistical models (Brown 1995; Rosenzweig 1995).

By the late 1970s, mainstream theoretical ecology and the search for biogeographical patterns of community structure had converged in two areas. One was related to species' body size. Mathematical models of competition and coevolution predicted that a species' body size should constrain its ability to coexist with other species. Competing species might show a striking regularity: there might be a limit to how similar in size species could be and still coexist (MacArthur 1970). This was qualitatively confirmed by the patterns in species' body size observed in many communities (Hutchinson 1959; Hutchinson and MacArthur 1959; Cody and Diamond 1975). The other was the development of island biogeography theory (MacArthur and Wilson 1967; Simberloff and Wilson 1969), based on the idea of dispersal limitation and local extinction of species. This theory was applied to understand community patterns on islands and fragmented habitats on continents (see Rosenzweig 1995). These areas of convergence suggested that further development of mathematical models of competition, coevolution, and colonization and extinction dynamics might produce a synthetic theory of community structure. Such a synthesis would be able to simultaneously explain the major patterns of species diversity and abundance, how these patterns incorporate the structure of species with different body size, and how they change with the scale of observation (see fig. 1.1 for some examples).

Synthesis would wait at least another twenty years. In the early 1980s, community ecology took a dramatically new turn. The predictions of theory had far outstripped experimental evidence from the

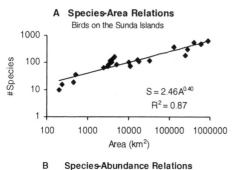

A   Species-Area Relations
Birds on the Sunda Islands

$S = 2.46A^{0.40}$
$R^2 = 0.87$

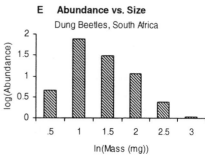

E   Abundance vs. Size
Dung Beetles, South Africa

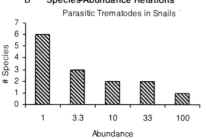

B   Species-Abundance Relations
Parasitic Trematodes in Snails

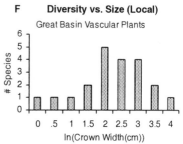

F   Diversity vs. Size (Local)
Great Basin Vascular Plants

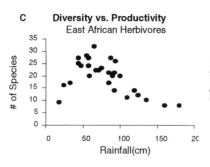

C   Diversity vs. Productivity
East African Herbivores

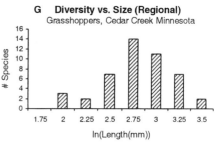

G   Diversity vs. Size (Regional)
Grasshoppers, Cedar Creek Minnesota

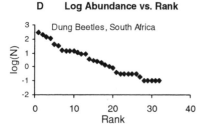

D   Log Abundance vs. Rank
Dung Beetles, South Africa

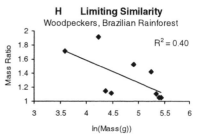

H   Limiting Similarity
Woodpeckers, Brazilian Rainforest

$R^2 = 0.40$

field. Although the population dynamical models of community structure could be used to interpret field data, alternative models that did
not invoke species interactions and evolution could predict some of the
same patterns. For example, body size patterns in communities were
sometimes equally well predicted by "null" models (Simberloff and
Boecklen 1979) of species with randomly assigned body sizes. This
continues today with models in which species traits do not determine
their abundance (Hubbell 2001; Harte et al. 2005, 2008). Moreover,
ecologists began to question whether communities were ubiquitously
structured by competition or contained populations anywhere near
population equilibrium (Lawton and Strong 1981). They argued that
other interactions, such as predation or mutualism, or mechanisms such
as colonization limitation and local extinction, might be just as important in structuring communities. These questions spawned a flood of
field experiments to determine the prevalence of competition, predation,
and other interactions within communities (Connell 1983; Schoener
1983; Sih et al. 1985). After two decades, these studies illuminated
complex food webs comprising dozens of direct species interactions
and even more indirect ones (Paine 1992; Wootton 1997). Communities appeared to defy any simple mathematical description, such as a

---

FIGURE 1.1. Examples of patterns in community structure that should be
able to be predicted by ecological theory from mechanisms of species coexistence, dispersal/colonization limitation, or neutral theory. (A) Species-
area curve for all bird species on the Sunda Islands (MacArthur and Wilson
1967). (B) Species richness-abundance relationship for trematode parasites
of the marine snail *Corithidia californica* (Lafferty et al. 1994). (C) Species
richness vs. rainfall, a surrogate of productivity, for mammalian herbivores
>300 g in 28 preserves in Kenya and Tanzania (Ritchie and Olff 1999). (D)
Log abundance vs. species rank in abundance for savanna dung beetles on
sandy soils (Coleoptera) in Mkuzi Preserve, South Africa (Doube 1991).
(E) Log abundance vs. body mass (mg) for savanna dung beetles on loamy
soils in Mkuzi Preserve, South Africa (Doube 1991). (F) Species richness of
vascular plants in the Utah Great Basin vs. size (maximum width of canopy/
stem) in a 10 x 10 m plot (Ritchie and Olff 2005). (G) Species richness of
grasshoppers (Orthoptera: Acrididae) found at Cedar Creek Natural History
Area, Minnesota (20 km^2) (Ritchie 2000). (H) Limiting similarity in a guild
of tropical forest woodpeckers, where mass ratio is the ratio of masses
(larger: smaller) of adjacent-sized species vs. the size of the larger species
in each pair (Terborgh et al. 1990).

community matrix of pairwise species interaction coefficients that
would determine the equilibrium abundances of species. Instead, they
seemed to require complex combinations of non-linear equations to
describe their dynamics (Abrams 1988; Schmitz 1992; Leibold 1996).
Even simple 3-species combinations had rich dynamics that exhibited cyclic or even chaotic behavior (May 1976; Huisman and Weissing 1999, 2001). The vision of predicting community structure and
its major patterns in any general way was all but abandoned by most
ecologists.

A FOUNDATION FOR SYNTHESIS

Despite the much-needed focus on field experiments over the past 20
years, many important conceptual developments during this period provide renewed optimism for a synthesis in community ecology. The application of hierarchy and complexity theory to ecology (Allen and
Starr 1982; O'Neill et al. 1986; Allen and Hoekstra 1992) suggests that
communities are too complex to be understood through the traditional
approach of using differential equation models for each species' population dynamics. Recent neutral models of community structure (Hubbell 2001; Bell 2001; Volkov et al. 2003; Alonso et al. 2006; Harte
et al. 2008) suggest that certain spatial patterns can result from simple
processes in space and time that do not depend on species traits. Ecologists also now increasingly recognize that many patterns in ecology
change with the scale at which they are observed (Greig-Smith 1983;
Wiens and Milne 1989; Levin 1992; Kunin 1998). Studies of interactions among plant species suggest that underlying trade-offs in the advantages of different physiological and morphological traits can explain
species coexistence (Grime 1979; Tilman 1988; Berendse et al. 1992;
Leibold 1989, 1996). New studies of how species' traits scale with
body size (Peters 1983; Calder 1984; Charnov 1994; West et al. 1997;
Enquist 2001) suggest that such trade-offs may depend generally on the
body size of organisms. Some experimental field studies of terrestrial
herbivores find that in fact species may be more likely to coexist than
expected because trait differences confer access to exclusive resources
(Schoener 1976; Belovsky 1986, 1997; Ritchie and Tilman 1993; Chase
1996; Ritchie 2002). These studies suggest that a species' ability to

coexist is driven more by the amount of its exclusive resources than by its competitive ability for overlapping or shared resources. Finally, the recognition that much of nature exhibits fractal geometry (Mandelbrot 1982) provides a potentially powerful tool for incorporating scale and heterogeneity into models of community structure (Morse et al. 1985). These seemingly disparate developments point to a potential revision in understanding community structure and diversity. Multiple differential equation models of consumer–resource interactions, one for each species, can generate the coexistence of many species (Huisman and Weissing 1999, 2001; Brose et al. 2004), but this approach seems unlikely to be useful in generating general predictions. The next best option perhaps is to focus on trade-offs in the advantages of species traits (Tilman 1990) and their potential for generating exclusive resources (Belovsky 1986, 1997; Ritchie 2002). Although there are many studies of trade-offs and coexistence (Sommer and Worm 2002), what is still missing is the connection of these traits to the distribution of resources in the environment, observed at different spatial scales (Levin 1992). More specifically, *how do species with different traits exploit biotic and environmental heterogeneity, how does this exploitation change with scale, and how does this allow coexistence?*

In this book, I develop a model to account for how species that are limited by the same resource might coexist. Specifically, I develop models to account for how community structure might be influenced by spatial and temporal heterogeneity in the distribution of habitat and resources across a range of scales of observation. The key assumptions of the model are:

1. potentially coexisting species all consume the same resource (MacArthur 1969; Tilman 1982);
2. individual organisms sample the environment at some characteristic scale or "grain" size that likely corresponds, on average, to a species' body size (Levins 1962, 1968; MacArthur 1972; Wiens and Milne 1989);
3. most organisms must consume some other material, which I call food, in order to acquire resources;
4. resources and habitats have a fractal-like heterogeneous distribution over some range of scales of observation (Milne 1992, 1997); and

5. individuals will select clusters of resources in a way that maximizes their fitness for a given distribution of resources and habitat (MacArthur and Pianka 1966; Emlen 1966; Schoener 1971; Charnov 1976; Stephens and Krebs 1986).

By applying these assumptions, some of which are based on some long-held ideas in ecology, to a classical model of the dynamics of a consumer species and its resource, I develop the mathematical background to describe the fundamental process of resource consumption in a heterogeneous environment. As I will show in the ensuing chapters, fractal geometry provides a "neutral" description of heterogeneity, that is, the pattern of heterogeneity is similar across scales of observation. However, *when fractal distributions are **sampled** with different grain sizes, the density and cluster size distributions of resources become explicitly dependent upon the scale of observation and measurement.* Emerging data (Johnson et al. 1996; Halley et al. 2004) suggest that fractal geometry provides a first-order approximate description of spatial distributions in nature that is certainly better than assuming random or uniform distributions (Tilman 1982; Tilman and Wedin 1991). David Morse and John Lawton (Morse et al. 1985) provided a hint of the potential of fractal geometry to address problems of resource availability and community structure (May 1988), and this idea has developed over the last decade (Milne et al. 1992; Palmer 1992; Ritchie 1997, 1998; Ritchie and Olff 1999; Olff and Ritchie 2002; Green et al. 2003; Halley et al. 2004).

The resulting model predicts the following. Species with different sampling scales (grain size) should be selective for different resource cluster (patch) sizes. An optimal minimum resource cluster size, or "giving up density," increases with sampling scale according to a simple scaling law. This minimum defines a niche boundary along a gradient of resource cluster sizes that ultimately determines whether a species can coexist with others. I test these predictions with results from several sets of data on minimum resource cluster sizes accepted by dung beetles, item sizes selected by mammalian and insect herbivores, and seed sizes and seed patch densities selected by desert granivores.

When resources are imbedded in another material (food: plant tissue, animal tissue, organic matter, water) that must be consumed, species with different sampling scales experience a trade-off in their optimal minimum food patch sizes and within-patch resource concentrations (Ritchie and Olff 1999). This trade-off depends fundamentally on heterogeneity, as it disappears when resources are random or homogeneous. This trade-off predicts the maximum, minimum, and optimal sampling scales for a guild of species that use the same resource, as well as their abundances and a limit to the similarity in sampling scales among species. Species of different sampling scales coexist because they have exclusive use of particular sets of food patch sizes and resource concentrations. To the extent that body size provides a first approximation to grain size, the model therefore predicts the number and abundance of species of different size that can coexist. Thus, the structure of a community can be predicted from the likely body sizes that would "sort" via competitive dynamics to allow the maximum number of species to persist. This predicted structure results independently of the detailed population and competitive dynamics of species, but is sensitive to environmental factors such as area, habitat fragmentation, and productivity. I then test the predicted patterns in species body size similarity, maximum and minimum body size, species' abundance, and species richness with data from many different guilds including mammals, birds, plants, arthropods, and marine zooplankton.

The resulting model yields a framework, based on population dynamics and evolution in a heterogeneous environment, for predicting major patterns of community structure. Because of its explicit inclusion of factors such as productivity, habitat distribution, and scale of observation, this framework can potentially be used to evaluate the consequences of environmental change, such as habitat loss, for community structure and biodiversity. As such it provides a niche-assembly-based model of community structure and species diversity that predicts multiple patterns that can be directly compared with those of alternative theories such as neutral theory (Hubbell 2001; Harte et al. 2008) or the "storage effect" (Chesson 1994, 2000).

I finish the book by placing this model in the context of other community structuring mechanisms such as colonization/extinction dynamics, disturbance, predation, and environmental constraints such as temperature. I also recognize that many large-scale patterns of community structure and diversity include all species in a taxon, regardless of resource guild. Clearly these patterns result from combining the structure of their subcomponent guilds, and thus from the "law of large numbers." However, because the model predicts the maximum, minimum, and optimum sizes of each guild, it can predict constraints on the diversity patterns for a collection of guilds. I explore these predictions and their support by field data to show how the traditional species patterns of community structure may emerge as the conglomeration of local, guild-specific patterns predicted by the deductive approach I present in this book.

This theory, if ultimately supported by field experiments and more data, furthers the original goals of the early community theorists, who perceived links between organism size, selective foraging, niche boundaries, and competitive coexistence. Because the model brings together several ideas in a novel way (scaling, fractals, exclusive resources, foraging for food rather than resources directly), few if any of its qualitative predictions have been tested experimentally. Thus, evidence from observed field data to support the predictions of the model is circumstantial for now. Other theoretical models, such as the neutral theory of biodiversity and biogeography (Hubbell 2001; Alonso et al. 2006) or spatial macroecology (Ostling et al. 2003; Green et al. 2003; Harte et al. 2005, 2008) suggest that trait differences among species do not determine community structure, and might in some cases explain observed data just as well, if not better, than classical niche models. Since mechanism can seldom be inferred from pattern, I suggest ways that the trait-dependent, niche assembly processes of resource competition might provide direct alternative predictions of the same types of patterns. The idea that multiple processes may simultaneously structure communities is not new (MacArthur and Wilson 1967; MacArthur 1972; Tilman 1994), but the development of a niche-based theory that predicts a full range of community patterns means that an explicit theory of combined niche and dispersal assembly may not be far away (Marquet et al. 2007).

This book in many ways rejuvenates a classic paradigm (Simberloff 1983): the composition, diversity, and abundance of species can be understood from their differences in morphology, behavior, and physiology (MacArthur 1972). Rather than "moving on" (Lawton 1999) to view community ecology as an amalgam of contingent and unpredictable interactions and processes (Simberloff 2004), this book provides optimism for community ecology. We know much more about how the morphology, behavior, and physiology of individuals influence inter-

actions with the environment and among species, as witnessed by liter-
ally hundreds of studies of species interactions and their modification
by the traits of species. The ideas inherent in the original formal development of niche theory and community structure in the 1960s and 1970s
are brought back in this book because they are very useful at synthe-
sizing the influence of species' traits on their interactions with the environment and other species. In particular, the book re-emphasizes
body size as an axis of niche differentiation and extends this concept
into an explicit theory of species' abundance and diversity. Moreover,
the incorporation of scale and heterogeneity, by using the new abstraction of fractal geometry, enlivens these classical "niche-assembly"
ideas so that they can explain many patterns of community structure in
unprecedented ways. The book is therefore far from a "dusting off" of
classics; rather it represents a new and potentially powerful solution to
a classical problem.

SUMMARY

1. Ecologists need to be able to predict multiple patterns in community structure from mechanisms that govern interactions among
   species in specific habitats across a wide range of spatial scales
   (experimental plots to continents).
2. I propose a new model for predicting the structure and diversity of
   ecological communities that incorporates two critical elements of
   diversity: scale and heterogeneity, into an analytically mathematical form.
3. As a first useful step, the new framework will use fractal geometry
   to simply describe heterogeneity in resource distributions in an explicitly scale-dependent way.
4. The resulting model will predict the existence and boundaries of
   exclusive niches for species with different sampling scales that use
   the same heterogeneous resource, and the limit to how similar such
   species can be and competitively coexist.
5. To the extent that sampling scale is associated with organism body
   size, these exclusive niche boundaries will also predict the relative
   abundance, maximum and minimum size, and diversity of species
   of different size.

6. The model's quantitative predictions will be compared with many different observed body size and diversity patterns for a wide variety of plant, invertebrate, and vertebrate communities.

7. Although the book will revisit some less popular but classical ideas of niche-based assembly of communities, the new abstraction of fractal geometry enlivens these ideas so that they can explain many patterns of community structure in unprecedented ways.

CHAPTER TWO

The Geometry of Heterogeneity

In this book, I explicitly incorporate spatial heterogeneity into mathematical models of niche boundaries, species coexistence and species richness. The first step is to define heterogeneity explicitly and quantitatively (Wiens et al. 1993). Dozens of studies now show that the physical environment, and the habitat and resources located within it, very often exhibit fractal-like geometry. This assumption generates the scale- or size-dependence of resource acquisition that produces most of the major qualitative predictions and quantitative tests in this book. In this chapter, I present a tutorial of fractal geometry to show how it captures heterogeneity and its scale-dependent properties. I also argue that there is abundant evidence that habitat and resource distributions are fractal-like over a wide range of scales of observation. Finally, I develop the fundamental mathematical scaling laws that incorporate fractal geometry, and thus spatial heterogeneity, into models of resource acquisition and habitat encounter that eventually produce predictions about community structure and diversity.

FUNDAMENTALS OF FRACTALS

If we look at the major physical objects in nature, such as clouds, trees, shorelines, rivers, mountains, etc., we find their shapes complex, irregular, or "rough." Unlike their abstract representations in art, such as in American folk art painting, these objects are not simple squares, circles, cubes, or spheres. Instead, they seem to be composed of copies of themselves. For example, from a small airplane flying toward a mountain range from a distance, the range would appear to be a single ridge. As the airplane draws closer, we find that this single ridge is composed of several smaller ridges that define the major drainages in the mountain range. Flying even closer, we find that these smaller ridges are themselves composed of ridges. If we continue flying into the mountain, we find that, just before we crash, the detailed topography of boulders, mounds, and swales mimics the pattern of ridges within ridges, within ridges, etc., of the entire mountain range. When objects show such similarity in pattern, or "statistical self-similarity" they are said to have fractal geometry.

As I show in this book, the rough, or fractal nature of the physical environment has profound implications for many aspects of organisms' existence, and thus their interaction and coexistence with other species at different spatial scales of observation. Why does it matter that the physical environment is fractal-like? Objects have rather non-intuitive properties. First, they do not form nice 1-, 2-, or 3-dimensional shapes of classical Euclidean geometry, but instead occupy fractions of these dimensions. For example, a map of a river network clearly occupies more than one dimension (a straight line) but also clearly occupies less than two dimensions (a filled plane). Second, objects with fractal geometry also have roughness because of their self-similarity: any given edge contains smaller versions of the edge. Viewing an object, like tree branches, at ever-smaller scales, we always find more branches, at least until we reach the level of leaves. This means that the measure of a fractal object cannot be objectively defined—it depends on the scale at which it is measured. This second property also yields a third property: as the scale of observation (not measurement!) increases, the total amount of Euclidean (1-, 2-, or 3-dimensional) space occupied by a fractal object declines. Put another way, the density of a given fractal material decreases as it is viewed at ever larger scales. As I will show in subsequent chapters, these properties of fractal distributions fundamentally affect the ability of species to acquire resources, encounter habitat, and coexist.

The bizarre properties of fractals are readily observed by examining a perfect mathematical fractal, such as a Cantor "carpet" (Mandelbrot 1982; Barnsley 1988) (fig. 2.1A). Such objects are perfect fractals because they are generated by precise mathematical rules that apply self-similarity at specific scales of observation. In the case of the Cantor carpet, a "generator" or underlying spatial pattern, can be "blown up" or represented as if it were part of a larger geometrical shape. For

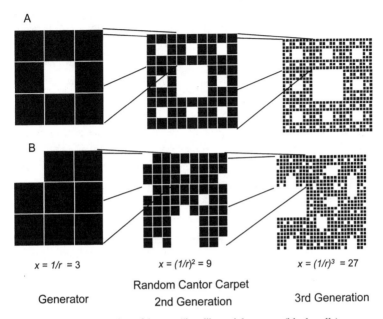

Deterministic Cantor Carpet

A

B

$x = 1/r = 3$ $x = (1/r)^2 = 9$ $x = (1/r)^3 = 27$

Random Cantor Carpet

Generator   2nd Generation   3rd Generation

FIGURE 2.1. Examples of how a "local" spatial pattern (black cells) can scale up in a self-similar fashion, producing a fractal-like spatial distribution over a finite range of scales. In this case a "generator" or local pattern is iterated across larger scales of observation to yield larger scale patterns that contain copies of themselves. (A) The exact or deterministic fractal of a Cantor carpet with 3 elements along one side or resolution $r = 1/3$, iterated twice. (B) A random Cantor carpet, in which the placement of occupied cells is randomized in each iteration of the generator. For more details, see Mandelbrot (1982).

example, suppose some process, such as the spread of fire on a landscape, the growth of leaves on a tree, or the growth of microbes on a surface, generates the geometrical shape of a square with the middle cell empty. If we expand the scale at which we observe the pattern, this square becomes one of many squares that surround a larger set of empty cells. If we iterate this expansion of scale of observation, then that set of squares becomes an empty centered square composed of smaller empty-centered squares. Notice that the carpet (black squares) does not fully occupy two dimensions and occupies proportionally less space as the scale of observation (the length of one side of the window or number of iterations) increases. These are properties 1 and 3 mentioned previously.

This example illustrates the fundamental properties of fractal geometry for pure mathematical fractals, but how do they relate to the types of spatial patterns observed in nature? Real objects often have a certain random element, with network branches occurring at random intervals or jutting off in random directions and large aggregations of material juxtaposed randomly next to smaller aggregations. However, this turns out not to be a problem. One can construct a randomized version of a pure fractal, in which the generator no longer has a precise geometric shape but instead is a randomly filled space (Mandelbrot 1982) (fig. 2.1B). If the amount of space filled in the random "generator," or smallest window size at which the pattern is manifested, is the same as in the pure fractal, then the random pattern is also fractal and in fact has the exact same properties as the pure fractal (box 2.1). Nevertheless, the appearance of the fractal is now very different, and its self-similarity becomes difficult to visualize. However, the spatial distribution looks much more like we would expect for real distributions relevant to biology.

A random Cantor carpet can be "tuned" to yield virtually any spatial pattern in 1, 2, or 3 dimensions by changing the length and proportional fill of the generator and the length of the window of observation. Such "random fractals," which simply assume that some pattern-forming process is repeated across a range of scales of observation, can generate very realistic-looking distributions. The local process acts as a random "generator," whose pattern is a self-similar subset of the overall object or distribution. Suppose that an object (or image) is subdivided into the smallest possible units (segments, cells, or cubes) of length ε. This length ε is the *scale of resolution* of the space we arbitrarily are investigating and is typically limited by the information available to the observer. The generator features N occupied units, arranged at random, of a length, area, or volume, respectively, equal to w^D where w is the number of units, of length ε, along one side of the generator, and D is the Euclidean dimension of the window of observation. For example, $D = 1$ if we consider a line, 2 if a plane, and 3 if a volume. We can increase the scale at which we iterate the generator by multiples of w, such that the extent of our window of observation x, is w^p, where p is an integer that refers to the number of times we have iterated the

BOX 2.1

Derivation of fundamental statistics for fractal-like

distributions of resources

Korcak's rule for number n_i of "islands" of material greater than or equal to a certain size z in any fractal-like distribution is $n_i = kz^{-b}$ where b is the ratio of the fractal dimension of the object of interest Q, to the dimension D of the landscape being observed (also known as the imbedding dimension (Mandelbrot 1982)). Since there are a maximum of k clusters in the entire distribution, the frequency of clusters greater than or equal to size z is just z^{-b}. It follows that, if cluster sizes differ by a constant ratio Δ (equivalent to differentiating cluster sizes on a logarithmic scale), then $f(z)$, the frequency of clusters of *exactly* size z, is the difference between the frequency of clusters greater than or equal to size z and the frequency of clusters greater than or equal to size Δz.

$$f(z) = z^{-b} - (\Delta z)^{-b}$$

$$f(z) = z^{-b}(1 - \Delta^{-b}).$$

For a given fractal-like distribution, the factor $(1 - \Delta^{-b})$ will be a constant c_z. Thus,

$$f(z) = c_z z^{-b}. \tag{2.1.1}$$

This derivation of $f(z)$ now allows calculation of the mean cluster size, which will turn out to be very important when we focus on how organisms sample fractal-like resources in the ensuing chapters. Following standard formulas for the moments of a fractal distribution (Milne 1997), the first moment (mean) amount of material per unit volume sampled,

$$\bar{G} = \int_1^{G_{max}} zf(z)dz.$$

Substituting for $f(z)$ yields

$$\bar{G} = \int_1^{G_{max}} zc_z z^{-b}dz, \; and$$

(Box 2.1 continued)

$$\bar{G} = c_z[G_{max}^{2-b} - 1]/(2-b), \tag{2.1.2}$$

where G_{max} is the maximum cluster size. If material fills the space, then G_{max} should be w^Q in the ideal case that the minimum number of volumes required to "cover" the distribution is sampled (Mandelbrot 1982; Barnsley 1988). Given that by definition, $w > 1$, such that $w^{Q(2-b)}$ is likely much greater than 1, the mean amount of material per occupied sampling volume (equation (2.9) in the text) is approximately

$$\bar{G} \cong c_z w^{Q(2-b)}/(2 - b). \tag{2.1.3}$$

Similarly, we can find the variance in the amount of material per sampling unit from the second moment of a fractal distribution (Milne 1997):

$$Var(G) = \int_1^{G\,max} \left(z f(z)\right)^2 dz \tag{2.1.4}$$

Substituting for $f(z)$ and G_{max} as above and evaluating the integral yields

$$Var(G) = [c_z^2/(3 - 2b)]w^{Q(3 - 2b)}. \tag{2.1.5}$$

Further analysis shows that the variance increases with the scale of sampling w, and unimodally increases and then decreases with increasing volume of resources, signified by Q.

generator (fig. 2.1A, B). Regardless of the *extent* of our window of observation x, the rate at which space is filled by the object is the same as the rate at which space is filled in the generator. Because the generator is iterated across scales of observation, the amount of space filled by the object or image follows a scaling law. By inference, the space-filling of the generator must obey the following scaling law (Mandelbrot 1982):

$$N = w^Q, \tag{2.1}$$

where Q is called the fractal dimension. If we know N and w, we can therefore solve for the fractal dimension by taking the logarithm of both sides of equation (2.1):

$$Q = \ln(N)/\ln(w). \tag{2.2}$$

Note that $Q \leq D$, since the object or image cannot fill more space than that in our window of observation. Thus, for a broken series of line segments, $Q < 1$, for a fragmented set of squares, $Q < 2$, and for a fragmented set of cubes, $Q < 3$. Equation (2.2) is fundamental to gauging whether real-world objects or images are fractal and estimating the fractal dimension (Voss 1986; Barnsley 1988; Milne 1992; Milne et al. 1992). The point is that it does not matter whether the N units of the generator are arranged in a geometric shape or randomly (fig. 2.1). How does this help us understand heterogeneity in nature? Visually, the final iteration of the random fractal in figure 2.1 begins to resemble the complex distribution of objects or materials in nature (Mandelbrot 1982; Barnsley 1988; Wolfram 2002). One can hypothesize that a given distribution of ecologically interesting objects, expressed in discrete units of a minimum length or resolution, is fractal-like. If so, then for a given scale of observation, or number of units of resolution equal to the length of the window of observation x, the object or image should fill a certain amount of space (fig. 2.2). This "mass" or amount of space-filled M, is determined by iterating the generator p times: $M = N^p = (w^Q)^p = w^{pQ}$. However, by definition, $w^p = x$, so

$$M = x^Q. \qquad (2.3)$$

Equation (2.3) is fundamental to measuring the fractal properties (if any) of natural objects or distributions of materials, and lays the foundation for determining the "mass" of habitat or resource a particular species might encounter (Milne et al. 1992; Ritchie 1997, 1998; Ritchie and Olff 1999; Pitt and Ritchie 2002). This relationship implies that occupied cells increase proportionally with the scale of observation, but at a slower rate than if the distribution has a Euclidean dimension D (1, 2, or 3). This property of fractals provides yet another way to assess whether a distribution is fractal-like and measure its fractal dimension, known as the sliding window method (Voss 1986; Milne 1992, 1997).

An alternative popular method, usually called "box-counting," of estimating "mass" of a fractal-like distribution is to measure the minimum number of sampling units (boxes) of length w required to cover the distribution. To get the minimum number of boxes, units must be centered on occupied cells. In this way, the boxes begin to approximate generators with random placement of occupied elements. In this

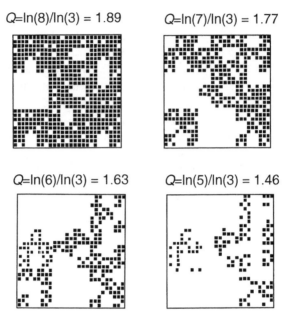

FIGURE 2.2. Examples of different random Cantor carpets with a generator of $r = 1/3$ iterated twice to yield different fractal dimensions Q by having different numbers of occupied cells N per generator, $Q = \ln(N)/\ln(1/r)$.

case, each sampling unit contains, on average, N or w^Q occupied elements (given eqn. (2.1)). If n_b is the number of boxes required to cover the object or image, then $n_b w^Q = M$, and

$$n_b = M w^{-Q}. \qquad (2.4)$$

This is the classic method of determining whether an image is fractal and estimating the fractal dimension (Barnsley 1988; Johnson et al. 1996; Milne 1997; Halley et al. 2004). More importantly, it shows the scale-dependence of the measurement of fractal objects and that the number of occupied sampling units declines with increasing sampling scale w.

The density, or proportion of our window of observation occupied, of a fractal object or image q, is directly related to space-filling, but in a way that we might not expect. The proportion of the window, of volume x^D, occupied is just

$$q = M/x^D = x^Q/x^D = x^{Q-D}. \tag{2.5}$$

This equation implies that, if the object or image is fractal-like with $Q < D$ (as I noted above), the proportional fill will get smaller as the scale of observation, x, gets larger. Equation (2.5) has an extremely important implication: *If x represents the scale at which the environment is observed, then the density of the distribution will decrease as the scale of observation or sampling increases* (Ritchie and Olff 1999; Haskell et al. 2002). Equation (2.5) also implies that Q and q are directly related to each other: $Q = D + \ln(q)/\ln(x)$. As q gets smaller, $\ln(q)$ gets more negative and the fractal dimension Q gets smaller.

The relationship between Q and q has another large implication: information about the *density* or proportional fill of a fractal-like object or image directly determines the geometry or *dimension* of that image. In turn, the geometry of the image directly determines the *scale-dependence* of any relevant measurements of the image. This connection between density, geometry, and the scale-dependence of measures is not widely appreciated in ecology, but has critical importance for ecological processes such as foraging for resources (Ritchie 1998; Haskell et al. 2002). As I will show throughout this book, this fundamental property of fractal-like distributions is key to understanding how ecological processes in space may be inherently scale-dependent.

Heterogeneity is often defined as the variance in density (mass) or other measure across space. Fractal geometry is in fact a "neutral" model of heterogeneity in the sense that the pattern in the distribution of substance is the same at all scales. Both the mean and variance of the amount of material in a fractal-like distribution, sampled with volumes (areas, segments) of length w, can be easily calculated if we know the relative frequency of clusters of different size. Korcak (1938) found that, in archipelagoes of islands, the number of islands $n_i(A)$ greater than or equal to area A obeyed the power law: $n_i(A) = kA^{-Q/2}$. The constant k is the total number of clusters of all sizes. The Korcak exponent is $Q/2$, where Q is the box-counting or mass fractal dimension of land in the archipelago, and 2 is the dimension D, of the environment in which the archipelago occurs. This relationship reflects a more general theorem, which I will call Korcak's rule (see proofs in Hastings and Sugihara 1993) that the number of clusters greater than or equal to size z in any fractal distribution are like islands in a landscape

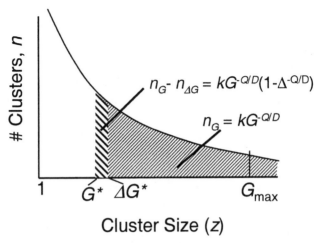

Cluster Size (z)

FIGURE 2.3. Hypothetical frequency distribution (shaded) of the number of clusters n of size z above a threshold size G, where k is the total number of clusters of any size in the landscape. A subset of this distribution is the number of clusters of exactly size G, or between size G and slightly larger clusters of size ΔG, $n_G - n_{\Delta G}$ (hatched area). This number of clusters yields $f(G) = (n_G - n_{\Delta G})/k$, or the probability of encountering a cluster of exactly size G. This generalizes to $f(z) = z^{-(Q/D)(1-D-Q/D)}$. Note that any given consumer will have a maximum cluster size G_{max} that they can encounter for any fractal distribution, given their sampling volume w^D.

of dimension D: $n_i(z) = kz^{-b}$, where b is the Korcak exponent Q/D (Hastings and Sugihara 1993). Since there are k clusters in the entire distribution (that is all clusters $\geq$ size 1; $n_i(1) = k(1)^{-b} = k$), the frequency of clusters greater than or equal to size z, $\phi(z)$, is

$$\phi(z) = n_i(z)/k = z^{-b}. \tag{2.7}$$

Box 2.1 shows how this analogy of clusters of material in space and "islands" in an archipelago generates approximate equations for the frequency of clusters of *exactly* size z $[f(z)]$ and mean cluster size $\bar{G}$ (fig. 2.3):

$$f(z) = c_z z^{-b}, \tag{2.8}$$

$$\bar{G} = c_z w^{Q(2-b)}/(2 - b). \tag{2.9}$$

These equations provide the mathematical foundation for how scale and heterogeneity are related for fractal-like distributions. Applied to resource distributions in subsequent chapters, they lay the foundation for incorporating scale and heterogeneity in classical ecological consumer-resource models and ultimately in models of species abundance, community structure, and species richness.

THE REAL WORLD: FRACTAL DISTRIBUTIONS OF RESOURCES

Calculating mathematical properties of fractals is relatively straightforward; the real question is how fractal is nature? More specifically, do organisms live in an environment where resources are distributed in a fractal-like manner? When the distributions of different habitats and resources have been measured, the answer is often yes.

Organisms occupy physical environments that, across 2–3 orders of magnitude in scale larger than the size of the organism, are very often fractal-like. Physical surfaces, such as land topography (mountain ranges, stream and river drainages, tidal flats, etc. (Turcotte 1995)) are commonly characterized as fractals. In streams, the surface of submerged rocks and cobble can be fractal-like (Cooper et al. 1997). In soil, particles are often heterogeneous in size and the interstitial spaces between particles where microbes and plant roots forage have fractal-like distributions over 2–3 orders of magnitude in scale (Bartoli et al. 1991; Rasiah and Aylmore 1998; Finlay and Fenchel 2001). The surface of vegetation also often has a fractal distribution, either from the combined architecture of individual plants (Morse et al. 1985; Gunnarsson 1992; West et al. 1999) (fig. 2.4), or in the distribution of plants across the landscape (fig. 2.5). Arthropods inhabiting the surfaces of lichens experience a fractal-like structure of folds and crevices (Shorrocks et al. 1991). Plant roots are themselves often fractal-like over 3 orders of magnitude in scale (Fitter and Stickland 1992). In pelagic, aquatic environments, where resources are usually considered well mixed, different resources, such as microbes, phytoplankton, zooplankton, and krill may be concentrated within complex temporary turbulent structures, such as eddies or upwelling currents (McClatchie

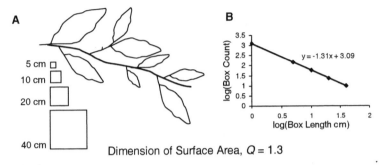

Dimension of Surface Area, $Q = 1.3$

FIGURE 2.4. Schematic of the leaf and stem architecture analyzed by Morse et al. (1985) to assess the surface area available to insects of different size. (A) Individual tree branches were placed flat on a table surface and the number of boxes of different lengths (cm) required to cover the leaves and stems were counted. (B) The change in log (box count) with log(box length) produces a highly linear response with a slope equal to $-Q$, where Q is the fractal dimension.

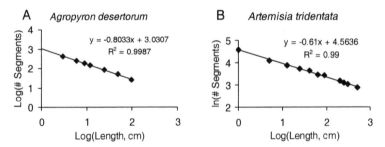

FIGURE 2.5. Box-counting results for the distribution of the dominant plant species in two different areas of sagebrush steppe. The graphs show how the number of segments of a 100 m line-intercept transect occupied by either (A) crested wheatgrass *Agropyron desertorum* and (B) big sagebrush *Artemisia tridentata* change with the length of segments used to subdivide the transect. Because the transect in each case is a line bisecting a 2-dimensional distribution, the slopes of each plot are equal to $1 - Q$, where Q is the fractal dimension (Voss 1986). The dimension of the grass ($Q = 1.80$) is higher than that of sagebrush ($Q = 1.61$), suggesting that sagebrush is more patchy and clustered.

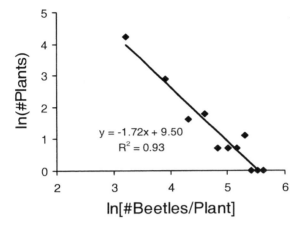

FIGURE 2.6. Scaling law frequency distribution of different numbers of insect herbivores living on individual stalks of goldenrod *Solidago* spp. (data from Root 1996), implying a fractal-like distribution of beetles across the landscape.

et al. 1994; Tsuda 1995; Lovejoy et al. 2001; Seuront and Lagadeuc 2001; Waters et al. 2003). From these examples, it also makes sense that habitats would also have fractal-like distributions. For example, the distribution of heathland habitat in 9 x 9 km regions of the Netherlands (fig. 2.6) is fractal-like over 2 orders of magnitude in scale of observation, even in areas where it is highly contiguous or very rare (Olff and Ritchie 2002).

If habitats have fractal distributions, then it would be logical to assume that the organisms that occupy those habitats might also have fractal distributions. For example, the prairie grass *Bouteloua gracilis* in New Mexico has a fractal distribution with an estimated dimension of 1.88 (Milne et al. 1992). Furthermore, combined pine species in the southern U.S. (*Pinus* sp.), and pinyon pine *Pinus edulis* in the southwestern United States exhibit fractal-like distributions across scales of 0.01 to 1000 km (Milne 1997). I found similar fractal-like spatial distributions of big sagebrush *Artemisia tridentata* in a Utah steppe (Ritchie et al. 1994) across a range of scales from 0.01 to 100 m (fig. 2.5). Being fractal over this range of scales means that the distribution pattern of plant material is the same for both architecture (stems, leaves) and in distribution across the landscape.

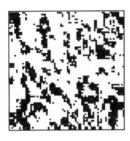

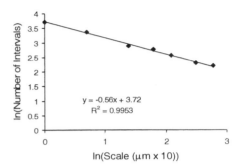

FIGURE 2.7. Fractal distribution of a biofilm of the bacteria *Myxococcus xanthus* (Welch and Kaiser 2001. The image is a micrograph filtered for highest contrast, such that black pixels are those more than 2 cell-layers thick. The interval-counting method reveals a highly significant scaling relationship of number of occupied intervals vs. interval length with a slope $-m$, where the fractal dimension $Q = 1 + m = 1.56$.

The fractal nature of animal distributions has not been explored in as many cases, but there are a few examples. Bald eagle nests are distributed as a fractal that mimics the fractal coastline of Admiralty Island, Maine (Milne 1997). As the previous examples show, plant distributions can also be fractal, and further evidence suggests these can extend to smaller (0.1–100 m) scales (Kunin 1998; Ostling et al. 2003; Green et al. 2003). Furthermore, the distribution of the copepod *Neocalanus cristatus* consists of multiple clusters of different size, and the two-dimensional distribution as sampled from the ocean surface (Tsuda 1995) is fractal-like with an estimated fractal dimension of 1.8 (±0.02 SE) over scales of 10–100,000 m. In yet another example, aggregations of beetles on a host plant (goldenrod, *Solidago altissima*) (Root 1996) have frequency distributions that are well described by a power law and thus are likely fractal (fig. 2.6). Similarly, cluster sizes of the nests of the neotropical ant *Azteca instabilis* follow a power law, implying a fractal distribution among otherwise uniformly distributed coffee trees (Vandermeer et al. 2008). Extending this to the smallest scales, bacteria organized as a biofilm on an agar plate, an environment virtually without structure, exhibit a fractal distribution (fig. 2.7, Welch and Kaiser 2001).

Animal movement paths also appear to have a fractal-like nature, perhaps in response to fractal-like distributions of resources or prey organisms. Organisms such as large ungulates (McClure et al. 2005), mammalian predators (Phillips et al. 2004), and insects (Wiens et al. 1993; With 1994) move in random walks, correlated to varying degrees (Turchin 1996), which are fractal-like over a range of scales of measurement (Hastings and Sugihara 1993). A recent review (Sims et al. 2008) shows that marine predators, including fish, sharks, and turtles, move in correlated random walks that mirror the fractal-like distribution of their prey. The dimension (degree of space-filling) and thus total space occupied by individual animals (area displacement) over time depends on the frequency of turns and their mean angle, but these factors do not change the fractal-like nature of the movement path and thus the implied fractal-like distribution of multiple individuals.

Of course not all resource or habitat distributions are fractal. Many plant species show very different distributions at small scales than at large (Hartley and Kunin 2003, Harte et al. 2005, 2008). Patterns can shift at certain scales, creating scale breaks (Allen and Holling 2002), in which the pattern at smaller scales differs from that at larger scales of resolution or observation. Processes such as territorial defense that govern local spacing behavior of animals can produce random distributions at small scales but aggregated fractal distributions at large scales (Schooley and Wiens 2001). Nevertheless, the large number of studies that have now found fractal-like physical habitats, vegetation, and distributions of individual organisms suggests that we can certainly accept fractal-like distributions of resources as a hypothesis. It makes little sense to focus *just* on whether nature *is* fractal-like over relevant ranges of scales, although this question still needs to be answered for many biological phenomena (Halley et al. 2004). Instead, if nature is often fractal, how does this affect the distribution and abundance of organisms and biodiversity, and the functioning of ecosystems (Palmer 1992; With and King 2004)? Fractal geometry is an approximate simplification, or even "null" model, of a complex world whose heterogeneity pattern is similar across scales and therefore is approximately described by a simple scaling relationship or power "law." *At the very least, fractal geometry offers a first-order approximation or description of the frequency of different cluster sizes and the rate at which substance accumulates with increasing scale of observation* (Milne 1997).

The purpose of this book in employing fractal geometry is to "see how far we can go" towards a theory of community structure that explicitly incorporates scale and heterogeneity and its interaction with species' body size. Heterogeneity is largely missing from models of consumer-resource and other species interactions in ecology, so the analytical tractability and "neutrality" of fractal geometry make it a good place to start. As I will show in ensuing chapters, a fractal description of complex distributions of resources and habitat offers fundamental insights about how organisms exploit resources, and helps us understand how they compete and coexist. In this regard, fractal geometry is a valuable tool.

SUMMARY

1. The complex distributions of habitat and resources in nature seem to be fractal-like, in that the same spatial pattern is repeated across a range of scales of observation.
2. Although most people are familiar with so-called deterministic fractals, in which a regular geometric pattern is repeated at increasing scales of observation, their mathematical properties are virtually identical to "random" fractals, such as a random Cantor carpet, which repeat random patterns in the distribution of material across scales. Random fractals produce patterns that look very similar to distributions in nature and can arise from almost any sort of spatial process.
3. Fractal-like distributions are metaphysical in a sense, in that measurement values depend on the scale at which they are measured.
4. This scale-dependent feature of fractal-like distributions means that the density of material (habitat, resources) declines as the scale of observation and scale of measurement increase. The dimension, or geometry, of fractal-like images is directly a function of the density of the image, and implies that any process (foraging, dispersal) that is sensitive to density will be scale-dependent.
5. A fractal-like distribution allows the frequency of different cluster sizes to be expressed as a simple scaling law of the scale of measurement (organism sampling scale), and this allows the direct calculation of the mean and variance in cluster sizes of material.

6. Physical surfaces (landscapes, soils, fluids) are very often fractal over 2–3 orders of magnitude in scale of observation, and accordingly, the distribution of vegetation and habitats is often fractal over similar ranges in scales. The distribution of organisms other than plants is less studied, but it seems logical to expect that they too may be fractal over limited ranges in scale.

7. Fractal geometry is a useful simplification to describe complex distributions in nature, and because of this simplicity, it is a critical tool in developing a theory of community structure that explicitly incorporates scale and heterogeneity.

CHAPTER THREE

Scaling Relationships for the Consumption of Resources

To persist in an environment, organisms must find enough resources to meet their requirements to produce enough offspring to replace themselves in their lifetime. They must avoid being eaten by other organisms, find mates (if they are sexual) and not experience intolerable physical conditions. If organisms are mobile, they can actively search the environment, while sedentary organisms must have resources delivered to them by the physical movement of the medium in which they live, such as water, or by the movement of potential prey. Other organisms, such as plants, search the environment by building permanent structures, such as plant roots, to deliver resources to the rest of the organism. In any of these scenarios, organisms consume resources at rates determined by the density and distribution of the resources. This rate, in the balance of other potentially limiting factors, must at least match resource requirements for individuals of a particular species to persist.

In this chapter, I explore how a heterogeneous supply of resources in space affects the rate at which individuals acquire resources and the resource availability at which a species can persist in the environment. To continue the theme developed in chapter 2, I use fractal geometry to describe the spatial distribution of resources when they are heterogeneously distributed in space. I show that the rate of resource acquisition depends strongly on the scale at which organisms sample the environment. This sampling scale, which may be highly correlated with body size, sets different threshold concentrations of resource density for species with different sampling scales or body sizes. Very often, vital resources are contained within some other substance, such as plant tissue, animal flesh, water, etc., that the organism must consume in order to obtain the resource. This substance, which I call "food," also may be heterogeneously distributed in space. If so, then species with different sampling scales have different requirements for food density and resource concentration within food. When the distributions of food and resources are fractal, these different requirements change with sampling scale as simple power functions with exponents of opposite sign. Food density requirements increase as the organisms' sampling scale increases, while resource concentration requirements decline. As I will show in later chapters, this trade-off fundamentally allows the coexistence of species that use the same resource and constrains the number and relative abundance of coexisting species.

CONSUMER-RESOURCE INTERACTIONS IN A HETEROGENEOUS ENVIRONMENT

As discussed in chapter 1, I propose that species' persistence, and thus their *niche* requirements, can ultimately be measured in terms of resources acquired. Individuals may die from being eaten by a predator, from disease, exposure, or starvation. However, they must acquire sufficient resources to grow and to survive long enough to at least replace themselves, on average, in their lifetime. Ultimately, the persistence of a particular species will be governed by the rate at which its individuals acquire resources relative to the rate at which they lose resources through metabolism or death (Schoener 1973, 1976). This is true regardless of the source of death. The impact of any resource loss or mortality factor therefore can be measured in terms of the resources necessary to compensate for losses (Leibold 1996, 1998; Chase and Leibold 2003; Murdoch et al. 2003).

As I discussed in chapter 1, the debate about the controls of species population sizes and community membership has revolved around the question of whether or not species' populations are at equilibrium maintained by density dependent population growth (and presumably limited resources). This debate overlooks the most important issue of classical niche theory (Hutchinson 1959; MacArthur and Levins 1964; Leibold 1996; Chase and Leibold 2003), which is simply whether species can persist under the resource availability, physical conditions,

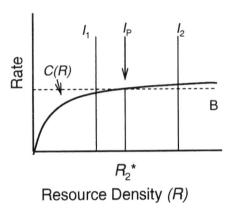

Resource Density *(R)*

FIGURE 3.1. Per capita consumption rate of a given species as a function of resource density R in the classic form of a Type II functional response $C(R)$, relative to the supply rate of resources in a resource-poor (I_1) and resource-rich (I_2) environment; and the per capita loss rate (for maintenance and replacement reproduction) of resources B (dashed line). With resource supply rate I_1, $C(R) < B$, and the species cannot persist. With resource supply rate I_2, the species' population will persist, grow and eventually reach an equilibrium population size when the resource density R is reduced until $R = R^*$. The species persists only if I exceeds a threshold rate I_P which also equals R^*.

natural enemies, etc., that exist in a given environment. Certainly, competition for resources may prevent a species persisting in a community (Leibold 1996), when density-dependent population growth results in species being near their equilibrium density. However, if predation, disturbance, or density independent mortality rates result in a high loss of individuals and/or unfavorable physical conditions cause high metabolic rates, species may be absent from a community because individuals cannot acquire resources fast enough to replace these losses (fig. 3.1). Thus, a species may be far from its equilibrium density but still be unable to exist in a particular environment. Nevertheless, the conditions that allow individuals of a species to exist can be measured in terms of the rate at which resources are supplied in the environment, or *resource supply rate* (Tilman 1982). This collapses the description of the species' niche into a "mass balance" equation of individual resource inputs and losses, as they are affected by the many factors that ultimately determine a species' niche (Leibold 1998; Chase and Leibold 2003).

To develop this idea rigorously, I use the familiar framework of consumer-resource interactions developed by MacArthur (1969), Tilman (1976, 1982), Abrams (1988), DeAngelis (1992), Leibold (1996), Chase (1999), Murdoch et al. (2003), and others. In this framework, resources are supplied to the environment at a certain rate I, which corresponds to a particular resource density R. The resource density determines the rate at which an individual of a particular species consumes resources, according to a function, $C(R)$, otherwise known as the functional response (Holling 1959). Resources are also lost at a per capita rate B through metabolism and death of individuals and converted into new consumers at efficiency a (Schoener 1973; Tilman 1982). These assumptions yield the familiar model of the change in density of resources (R) and number of consumers (N) over time:

$$dR/dt = I - C(R)N,$$

$$dN/dt = Na[C(R) - B].$$ (3.1)

If $C(R)$ at resource supply rate I exceeds B, then the species' population will grow, reducing the available resource density R. As this resource reduction proceeds, the per capita population growth rate will decline with increasing N, that is, *in a density dependent* fashion. The population will reach equilibrium when $C(R) = B$, and R is reduced to the value R^* (Tilman 1976, 1982). This basic model lies at the heart of community ecology based on consumer-resource interactions (Tilman 1982, 1988; Grover 1990; Ritchie and Tilman 1992, 1993; Leibold 1996, 1998; Chase and Leibold 2003; Murdoch et al. 2003). However, sometimes I may not yield a $C(R)$ fast enough to exceed B. In this case, resource supply and the factors contributing to resource loss (unfavorable physical conditions, predation, disease, disturbance, etc.) prevent the species' persistence. Most importantly, the threshold resource supply rate for simple persistence I_p is equal to R^*, the equilibrium resource density. This means that the resource availability conditions for one individual to persist are the same as when a population reaches equilibrium. Interpreted in terms of niche theory (Leibold 1998; Chase and Leibold 2003), I_p and R^* represent a niche "boundary" for a given species.

This classic consumer-resource model applies if resources are "well-mixed," that is, randomly or uniformly distributed. Under this assumption, every unit of resource can be used by every species. In experimental studies, this assumption has been met for mineral nutrients in chemostats for microbes, phytoplankton and zooplankton (Tilman 1976; Hsu et al. 1977; Rothhaupt 1988) and in purposely well-mixed garden soils for perennial grasses (Tilman and Wedin 1991). This assumption, however, is not met in other studies, such as phytoplankton gathering light (Huisman and Weissing 1994) or herbivorous insects feeding on vascular plants (Ritchie and Tilman 1992, 1993), where resources were clearly supplied unequally in space. These results suggest that ecologists still know very little about the consequences of spatial heterogeneity in resources for consumer-resource interactions, and, as I will discuss later, community structure and biodiversity.

If resources are supplied across space in a fractal-like distribution, as I argue in chapter 2, then the effect of spatial heterogeneity in resource supply on the ability of individuals to acquire resources can be explored by modifying the classical model (eqn. (3.1)). More specifically, the functional response $C(R)$ must now account explicitly for fractal rather than randomly distributed resources, as is assumed in the classical model. This means that there is a probability distribution of the "local" density of resources encountered within any sampling volume of length w and dimension D within a "landscape" of extent x and dimension D. The fractal geometry relationships developed in chapter 2, specifically equations (2.7)–(2.9), can be used to characterize this distribution of local resource densities, or resource cluster sizes, detected by an individual consumer:

$$dR/dt = I - C(q,Q,w)N \text{ and}$$

$$dN/dt = Na[C(q,Q,w) - B], \qquad (3.2)$$

where q is the proportion of landscape occupied by resource. I assume that q is measured at a scale of resolution, ε, by an observer and represents the proportion of pixels of volume ε^D in a D-dimensional landscape consisting of x^D pixels. Q is the fractal dimension of resources and w is the length of the sampling unit used by the consumer species. All scales, including the sampling scale w, are measured in units of ε, so $w = 1$ would mean that w is equal to the scale of resolution used by the observer (fig. 3.2), while $w = 8$ would mean that the organism samples an area 8 times ε in length.

Extent

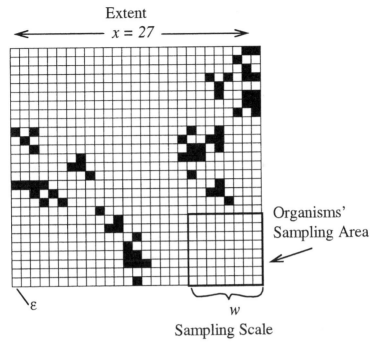

Sampling Scale

FIGURE 3.2. Hypothetical 2-dimensional landscape ($D = 2$) generated from a random Cantor carpet showing a fractal-like distribution of resources (black pixels) with fractal dimension $F = 1.27$. The landscape is characterized by the observer, who defines the presence/absence of resources in pixels on a map with resolution ε (length of one pixel). The organisms' sampling scale, w, is then measured as multiples of units of resolution. In this case $w = 8$ or 8ε. The extent of the landscape x, measured in units of resolution, is 27.

The function $C(q,Q,w)$ implicitly incorporates information about the resource abundance and distribution, as perceived by individuals of the consumer species. This general function implies some important insights. First, the consumer cannot be present in the environment unless $C(q,Q,w) \geq B$ when $N = 1$. Thus, the consumer's sampling scale w determines the resource abundance and distribution, defined by q and Q, at which a species can exist in the environment. Second this condition for persistence explicitly states the niche boundary of a species because consumer species with certain sampling scales should not be able to persist at sufficiently low supply rates, and these "threshold" supply rates may differ for species with different sampling scales. Third, resource distribution, as described by the fractal dimension Q, and organism scale w are directly related to species' persistence.

FORAGING IN A FRACTAL ENVIRONMENT

Resource consumption $C(q,Q,w)$ is determined by the movement of a consumer through a landscape with a fractal-like distribution of resources and its encounter with different "local" resource densities. A consumer will sample a volume V over the time period dt over which consumer and resource densities change. In that volume, the consumer will encounter resource in any given sampling volume with probability p_E. If resources are encountered, the consumer will find a range of resource cluster sizes G that yield resource densities G/w^D (dividing cluster size by sampling volume). I also assume that each cluster requires a time h (handling time) to process that is independent of cluster size. Each cluster size occurs with probability $f(G)$. Therefore,

$$C(w,q,Q) = Vp_E \frac{\displaystyle\int_{G\min}^{G\max} Gw^{-D}f(G)dG}{1 + hVp_E \displaystyle\int_{G\min}^{G\max} f(G)dG}. \tag{3.3}$$

Interestingly, consumers might acquire resources at no time cost by direct diffusion or uptake from the environment, such as the absorption of nutrients from the water column by algae. If so, the denominator of equation (3.3) disappears and, as I show below, the consumer should simply take any resource wherever encountered, as assumed by classical consumer-resource interactions (MacArthur 1970; Tilman 1982; Chase and Leibold 2003).

If resources are fractal-like in their distribution, this implies that the fundamental scaling relationships that describe fractal distributions yield general equations for V, p_E, G_{max}, and $f(G)$ (Milne 1992; Milne et al. 1992; Ritchie 1998; Olff and Ritchie 2002, see chapter 2). In particular, being able to specify a general relationship for $f(G)$ allows a general optimal foraging equation for selection of an optimal minimum resource cluster size, G_{min}, to be derived (Ritchie 1998). Fractal geometry therefore allows optimal foraging to be explicitly connected to consumer-resource dynamics in a way that will potentially link the selection of different resource clusters to the abundance and persistence of species. This also emphasizes how optimal foraging theory may reflect how organisms exploit heterogeneity in a way even more general than that originally conceived by its progenitors (MacArthur and Pianka 1966; Emlen 1966; Schoener 1971; Pulliam 1974). Imagine a foraging organism, moving randomly through a landscape of extent x and dimension D (fig. 3.2). As outlined in chapter 2, the dimension D is usually 1, 2, or 3, a Euclidean dimension imposed by the observer. The extent x is the number of pixels along one length of the landscape. The resource is distributed with fractal dimension Q and occupies a fraction q of the pixels in the landscape. In the example landscape (fig. 3.2) characterized as a plane ($D = 2$) with 27 pixels along one side and 60 pixels occupied by resource would have extent $x = 27$, and resource density $q = 60/729 \cong 0.0823$. The fractal dimension Q will depend on the number of occupied cells in the generator. Since this is usually unknown, Q can be estimated by statistical methods, such as box counting, outlined in chapter 2.

At any given time, the consumer samples some space of length w (measured in number of units of resolution) and volume w^D. I refer to volume, because organisms can use space in 3 dimensions. However, it is much easier to illustrate the fundamental concepts in two dimensions. For example, in the example landscape (fig. 3.2), $w = 8$ and volume $w^D = w^2 = 64$. By moving, the consumer samples v volumes over the time interval dt. Alternatively, vw^D can be the volume of flowing medium sampled by a sit-and-wait consumer per unit time, such as a caddisfly lodged on a rock in a flowing stream. Thus, the total volume in the landscape sampled is $V = vw^D$. For example, if the organism in the hypothetical landscape (fig. 3.2) sampled 4 volumes per unit time, then the total volume sampled would be 256 units.

The consumer has some probability p_E of encountering resource in a given sampling volume. This probability is determined as $1 - p_O$, where p_O is the probability that all the w^D cells in a sampling volume are empty. Individual cells are occupied with a probability q or the proportion of the landscape occupied by resource, so p_O is the joint probability that all w^D cells of a sampling volume are empty. Thus

$$p_o = (1-q)^{w^D},$$

and thus

$$p_E = 1 - (1 - q)^{w^D}.$$

Going further, we remember from chapter 2 that q depends on the extent of observation x in that $q = x^{F-D}$. Thus,

$$p_E = 1 - (1 - x^{Q-D})^{w^D} \tag{3.4}$$

This result makes a major qualitative prediction: *the probability of encountering fractally distributed resources during random movement through the environment depends on both the scale of observation and the sampling scale of the consumer.* However, as w increases, the probability of encountering resource in a sampling volume of length w, or p_E, quickly saturates to 1, unless we are studying the consumer across very large scales ($x > 10^9$) and resources are extremely rare (<100 ppm, $Q < 1$). The situation for large landscapes and/or very small q is revisited in chapter 9, where a $p_E \ll 1$ results in the possibility that resources clusters are not encountered by individual consumers and thus a form of colonization limitation occurs. My purpose for now is to explore how individual consumers respond to local conditions (small x), so I will assume for now that $p_E = 1$. Note that, if the resource is randomly distributed such that $D = Q$, then p_E is always 1, regardless of x or w, because $D - Q = 0$.

In each occupied sampling volume, the amount of resource present, G, in any given sampling volume will vary, given that foraging organisms are unlikely to visit locations in a way that precisely minimizes the number of sampling volumes required to "cover" pixels occupied by resources (Turchin 1996). Consumers may be selective and actually consume resources in a sampling volume only when their density exceeds some threshold (Stephens and Krebs 1986; Ritchie 1998). This selectivity implies that consumers might consume resources only in sampling volumes with G_{min} or more resources. Therefore, the amount of resource consumed per unit of occupied sampling volume is G multiplied by the probability, $f(G)$, that if resources occupy the sampling volume, resources occur in amount G.

The frequency of different cluster sizes $f(G)$ is determined by another inherent property of fractal distributions that I reviewed in chapter 2: Korcak's rule (box 2.1). Applied to resource distributions, the frequency of cluster size $G f(G) = c_G G^{-b}$ where the Korcak exponent $b = Q/D$, and the constant $c_G = 1 - \Delta^{-b}$ where again, Δ is an arbitrarily chosen ratio of larger to smaller cluster sizes of a given pair in an ordered set of cluster sizes. In chapter 2, I showed that "covering" a fractal map of dimension Q with the least possible number of sampling volumes yields w^Q material per sampling volume. Therefore, an optimal placement of a sampling unit on a resource distribution will encounter at most w^Q resources per sampling unit, so $G_{max} = w^Q$. Thus a randomly moving consumer with sampling scale w can encounter cluster densities from 1 to w^D in occupied sampling volumes. The issue is then, which of these resource clusters will it consume and which will it pass by in order to maximize its rate of resource consumption.

Now that we have derived explicit scaling relationships for V, and $f(G)$ and G_{max}, and have assumed that the probability of encounter, $p_E = 1$, we can substitute these into equation (3.3) to obtain the consumption rate of resources as a function of the scale of the consumer and the distribution of resources.

$$c(v,w,Q) = \frac{v \cdot c_G \cdot w^D \cdot \int_{G_{min}}^{G_{max}} G^{1-b} w^{-D} dG}{1 + h \cdot v \cdot c_G \cdot w^D \cdot \int_{G_{min}}^{G_{max}} G^{-b} \cdot w^{-D} dG}. \tag{3.5}$$

Simplifying and evaluating the integrals yields:

$$c(q,Q,w) = \frac{\dfrac{v \cdot c_G}{2-b} \left[\left(w^Q \right)^{2-b} - \left(G_{min} \right)^{2-b} \right]}{\left[1 + \dfrac{h \cdot v \cdot c_G}{1-b} \cdot \left[\left(w^Q \right)^{1-b} - \left(G_{min} \right)^{1-b} \right] \right]}. \tag{3.6}$$

The time spent processing small clusters acts as an "opportunity cost" in the classic economic sense applied to optimal foraging (Stephens and Krebs 1986), that is, it is time that would more profitably be spent finding larger clusters. To avoid this cost, the consumer will consume resources only in sampling volumes with cluster sizes $>G_{min}$. As G_{min} approaches G_{max}, however, the consumer's total resource intake declines because such large clusters are increasingly rare. I can now find an optimal G_{min} that balances the cost of too poor a reward per sampling volume with too infrequent an encounter with a useable

BOX 3.1

OPTIMAL MINIMUM RESOURCE CLUSTER SIZE AND CONSUMPTION RATE,

AND MINIMUM RESOURCE DENSITY AND DISTRIBUTION

Beginning with equation (3.6) for resource consumption rate:

$$C(q,Q,w) = \frac{\dfrac{v \cdot c}{2-b} \cdot \left[(W^Q)^{2-b} - (G_{min})^{2-b} \right]}{\left[1 + \dfrac{h \cdot v \cdot c}{1-b} \cdot \left[(W^Q)^{1-b} - (G_{min})^{1-b} \right] \right]}, \qquad (3.1.1)$$

(where c is the constant c_G). Taking the derivative with respect to G_{min}, and setting it equal to zero yields the following polynomial:

$$0 = v \cdot c_G \cdot h \left(w^Q \right)^{(2-b)} - \left[(2-b) + \frac{h \cdot v \cdot c G}{1-b} \cdot \left(w^Q \right)^{(1-b)} \right] \cdot G_{min}$$

$$+ \leq \left(\frac{h \cdot v \cdot c G}{1-b} \right) \cdot \left(G_{min} \right)^{2-b} \qquad (3.1.2)$$

This polynomial cannot be solved explicitly, but the G_{min} at which the derivative of $C(q,Q,w)$ is zero is determined primarily by the lower order terms for G_{min}. Furthermore, we can assume that the scale of resolution at which we have information about the environment is small relative to the sampling scale w of the consumer, so that $w \gg 1$. With these two assumptions, we can ignore the higher order term containing G_{min}^{2-b} and the coefficient term $(2-b)$, and obtain an approximate optimal $G_{min} = G^*$ with a little algebra:

$$G^* = (1-b)w^Q. \qquad (3.1.3)$$

resource cluster. In the standard fashion, I take the derivative of $C(q,Q,w)$ with respect to G_{min}, set it equal to zero, and solve for G_{min} (box 3.1). This yields a simple scaling relationship for G^*:

$$G^* = (1-b)w^Q. \qquad (3.7)$$

This result for a minimum resource cluster size has a number of interesting implications. First, as resources become more rare and/or

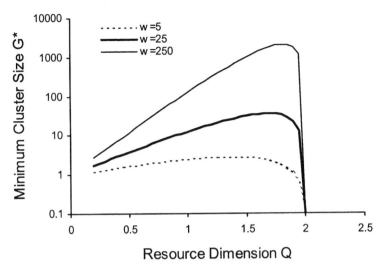

FIGURE 3.3. Change in the minimum resource cluster size that maximizes resource intake (G^*) as a function of increasing resource density and/or dispersion, that is, the fractal dimension (Q), for consumers with different sampling scales (w). Simulations of eqn. (3.7) with $D = 2$.

clustered (lower Q), the Korcak exponent b (which equals Q/D) decreases, such that G^* first increases and then decreases with increasing Q until resources become randomly distributed ($Q = D$), in which case consumers are completely unselective (fig. 3.3). Indeed there is an intermediate Q' at which G^* is maximized for a consumer with a given sampling scale w: $Q' = D - 1/(\ln(w))$. At very low $Q < 1$, resources exist primarily as single occupied cells, with little variation in cluster size, and thus there is little opportunity cost for consuming a resource. At very high Q (near D) where b approaches 1 and resources become very nearly randomly distributed, G^* approaches 0 because all cluster sizes tend to be large and there is again little opportunity cost for consuming a given cluster size. At these extremes, consumers with different sampling scales (w) will optimally select similar-sized clusters of resources. In contrast, at intermediate Q, a large range of cluster sizes are available, so the consumer may experience significant opportunity costs by selecting resource clusters that are too small. Consumers with different sampling scales will optimally select different-sized clusters of resources, and the minimum cluster size increases exponentially with increasing sampling scale w (fig. 3.3).

Another important result of equation (3.7) is that, for a given landscape, smaller-scaled consumers will include smaller resource clusters than larger-scaled consumers. This implies that the ratio of the minimum resource cluster sizes for two consumers with different sampling volumes will always be proportional to the ratio of their sampling volumes, regardless of resource density and distribution. Stated another way, the difference in the logarithms of G^* for two such consumers will always be proportional to the difference in the logarithms of their sampling volumes. This implies that, *at all possible resource densities and distributions, the selection of resource clusters will follow a scaling law*. This has profound consequences for the emergence of size structure within communities that I will explore in chapter 5. As I discussed in the previous paragraph, however, these differences in G^* may be very small when resources are very rare, or when Q approaches D, as in the case of randomly or uniformly distributed resources.

SCALE DEPENDENCE OF THE RESOURCE NICHE

Equation (3.7) directly links the distribution of resources to the behavior of individual consumers, and I can now use this link to show how the distribution of resources influences interactions between a consumer and its resource, and ultimately the consumer's resource-based niche (Chase and Leibold 2003). The optimal minimum resource cluster size G^* provides the maximum resource consumption rate for a given resource abundance and distribution, as defined by Q. The consumption of resource by the consumer of course influences the resource density and distribution, and therefore Q.

To find the niche boundary, or resource supply rate that allows persistence, I first derive an equation for the optimal resource consumption rate, C^* as a function of Q, given the optimal minimum cluster size G^*. I can substitute G^* from equation (3.7) for G_{min} in equation (3.6) and then simplify, which yields a complicated-looking expression (see eqn. (3.2.1) in box 3.2). At the low resource densities typical at persistence (or equilibrium density) of the consumer population, (see box 3.2 for the mathematical details), I obtain

---

BOX 3.2

OPTIMAL CONSUMPTION RATE AND RESOURCE DENSITY AND
DISTRIBUTION AT A CONSUMER'S NICHE BOUNDARY

The optimal minimum cluster size G^* can be substituted for G_{min} in
equation (3.1.1) (box 3.1) to obtain an optimal consumption rate C^*,

$$C^* = \frac{\dfrac{c \cdot v}{2-b} \cdot w^{Q \cdot (2-b)} \cdot \left[1-(1-b)^{2-b}\right]}{1+\dfrac{h \cdot c \cdot v}{1-b} \cdot w^{Q \cdot (1-b)} \cdot \left[1-(1-b)^{1-b}\right]}. \tag{3.2.1}$$

Simulation reveals that the term $[1 - [(1 - b)^{2-b}]/(2 - b)]$ is approximately b. At low, non-saturating resource densities, where handling time
costs are negligible, we obtain:

$$C^* = c_G vbw^{Q(2-b)}. \tag{3.2.2}$$

At the niche boundary, non-saturated $C^* \geq B$. This yields the following
equation, after recalling that $b = Q/D$, which can be solved numerically
for Q^*:

$$c_G v(Q^*/D)\, w^{Q^*(2-b)} = B. \tag{3.2.3}$$

$$C^* = c_G vbw^{Q(2-b)}. \tag{3.8}$$

This relatively simple equation has important implications for the
scale-dependence of a consumer's resource use and niche. First, *optimal resource consumption, like the minimum cluster size G^*, obeys a
scaling law* such that $C^* \propto w^Q$. This means that larger-scaled consumers have a much higher resource consumption rate than smaller-scaled
ones because they search a greater volume of space per unit time and
don't pay the time (opportunity) cost of consuming frequent small resource clusters. In addition smaller-scaled consumers are likely to encounter large regions of empty space and experience longer distances
and times between occupied sampling units, even though they detect
higher resource densities once resources are detected (fig. 3.4).

A second implication is that the scale dependence of consumption
rate implies a scale-dependence in the necessary resource density and
supply rate for persistence of a population of consumers. We observed

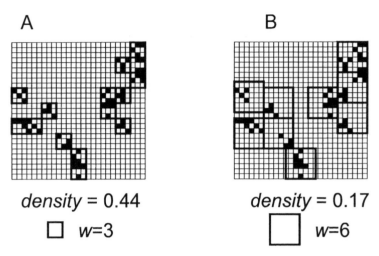

FIGURE 3.4. Consequences of sampling the hypothetical fractal-like distribution of resources (black) in fig. 3.2 at different scales. Sampling units are indicated by bold squares. (A) Average density detected by a smaller-scale forager ($w = 3$). (B) Average density detected by a forager with twice as large a scale ($w = 6$). Note that the smaller-scale forager requires 15 units to "cover" the resources, not counting many empty sampling units encountered while moving, while the larger-scaled forager requires only 9 units, with only 1 empty sampling unit in between otherwise connected units.

earlier in this chapter (following eqn. (3.2)) that the supply rate of resources must be high enough to exceed resource losses for a species with a given sampling scale w to persist, that is

$$C^* \geq B. \tag{3.9}$$

Furthermore, I showed that the resource supply rate for persistence, I_p, corresponds to a proportion of landscape supplied by resources q^* and fractal dimension Q^* equal to the resource density R^* at population equilibrium (eqn. (3.1)). Since C^* is scale dependent, then the minimum possible supply rate I_p that allows persistence will also be scale-dependent. Substituting equation (3.8) for C^* (assuming that resources are scarce at population equilibrium or at the threshold for species' persistence) into equation (3.9) yields equation (3.10), the condition for persistence of a consumer population (see box 3.2, eqn. (3.2.3)):

$$c_G v(Q^*/D) \, w^{Q^*(2-Q^*/D)} = B. \tag{3.10}$$

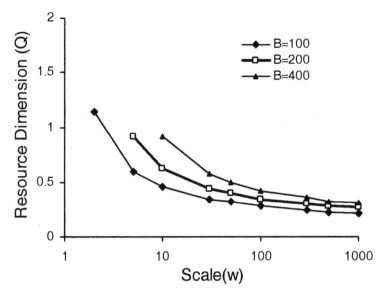

FIGURE 3.5. Simulations of the threshold fractal dimension of resources Q^* that allow persistence of consumers with different sampling scales (w) for 3 different resource requirements. Note the asymptotic-like behavior of the curves, suggesting that Q^* is relatively insensitive to sampling scale. Note also that the largest w has the lowest Q^* and thus can exist at the lowest resource density R^*.

As I showed in chapter 2, the fractal dimension Q is related to the proportion of landscape filled by resources and the ratio of the size of the "generator" (in this case the sampling scale w) to the extent, or scale of observation, x. This implies that there is a threshold, landscape-scale q^* below which a consumer with a particular sampling scale cannot persist. Following the math outlined in chapter 2 (eqn. (2.5)), $q^* = x^{Q^*-D}$. We can numerically solve equation (3.10) for Q^* (fig. 3.5).

The distribution of resources, which in classical models is assumed to be unaffected by consumers, is predicted to become increasingly rare but aggregated, as reflected by decreasing Q^*, with increasing foraging scale of the consumer (fig. 3.5). This means that consumers with different foraging scales may potentially produce different patterns of resource distribution at population equilibrium. Because of their higher minimum cluster size, larger-scaled consumers may ignore even reasonably clustered resources while smaller-scaled consumers may consume even small resource clusters, leaving only widespread, small clusters occurring across the landscape.

The threshold $q*$, the proportion of the landscape occupied by resource at persistence or population equilibrium, is analogous to $R*$ in the classical consumer-resource models of MacArthur (1969) and Tilman (1982). Given the positive relationship between q and Q, equation (3.10) and figure 3.5 suggest that, as expected from classical consumer-resource models, equilibrium $Q*$ (and thus $q*$) increases with the per capita resource loss rate B and decreases with increasing search effort v, the number of volumes sampled per unit time by the consumer. However, larger-scaled consumers consume resources at a faster rate but leave larger clusters of resources behind. Under the (perhaps dubious) assumption that different-scaled consumers have the same resource requirements B, larger-scaled species can persist at lower $q*$ and $Q*$ than smaller-scaled species. This is exactly equivalent to larger-scaled consumers having a lower $R*$ for the limiting resource than smaller-scaled consumers. Consequently, larger-scaled consumer species would be expected to competitively exclude smaller-scaled ones. Alternatively, resource requirements, B, might increase with sampling scale according to a scaling law (Peters 1983; Calder 1984; West et al. 1997), in which case $Q*$ might increase with sampling scale, rendering the smaller-scaled consumer as competitively superior.

Further inspection of the numerical solutions to equation (3.10) (fig. 3.5) however, suggest that the $Q*$ required for persistence does not change by more than 0.1 across >2 orders of magnitude in sampling scale. This insensitivity in the equilibrium or persistence density and distribution of resources to sampling scale is driven strongly by the fact that larger-scaled consumers are more selective of resource clusters and will tend to leave a larger fraction of resource *clusters* unconsumed. Nevertheless they are expected to search a much larger volume of space and can persist on lower overall *densities* of resource. This trade-off for larger-scaled consumers suggests that consumers with very different sampling scales may have approximately equal impacts on overall resource densities and distributions. This observation leads to the most interesting consequence of equations (3.7)–(3.10). They imply that *niche boundaries for consumers are at least as sensitive to their spectrum of resource use (cluster sizes) as to their ability* *to reduce resources to a lower overall density (or altered distribution)*. This prediction contrasts sharply with that of classical consumer-resource models, which focus only on impacts on resource density. The importance of resource selection will play an important role in the coexistence of consumers with similar foraging scales that I will explore in later chapters.

BODY SIZE AND FORAGING IN
HETEROGENEOUS ENVIRONMENTS

This modified consumer-resource model explicitly links the scale of the observer (x) and the scale of the organism (w) with the scale and geometry of the resource (resolution ε, and fractal dimension Q). The model provides a framework for incorporating resource heterogeneity into the dynamics of consumers and resources. The above analysis illustrates how heterogeneity in resource distribution alters the dynamics of consumers and resources from those predicted by classical consumer-resource models. The model emphasizes the dependence of species' foraging and persistence on both the sampling scale w and the distribution and abundance of the resource Q. Since sampling scale is likely to vary across a range of species with different body size, I will now use the modeling framework based on fractal geometry to address questions about how species of different body size acquire resources.

Throughout this book, I will assume that the sampling scale of organisms is proportional to their body length, as a first approximation. The actual sampling scale of organisms will depend on their morphology and sensory perceptions. Clearly, organisms that are an order of magnitude larger in size than others simultaneously sample a much larger volume of space and perceive the environment as more *coarse-grained* (Levins 1962). Other traits besides size, per se, may determine a species' grain or sampling scale, such as activity range, sensory system, etc. For example, a terrestrial herbivorous snail is unlikely to have the same sampling scale as a grasshopper (Orthoptera) even though they can be of similar body length and potentially consume the same food (Hulme 1994, 1996; Ritchie and Tilman 1992). Likewise, snakes, which rely on olfactory perception to hunt small mammal prey, may have smaller sampling scales than birds of roughly similar size, which fly and use vision to hunt similar prey (Capizzi and Luiseli 1996). However, across orders of magnitude in body mass for a similar trophic group or "guild" (e.g., granivores, sap-feeders, leaf-eaters, carnivores of vertebrates, etc.), body size is likely to be the best predictor of sampling scale, or the scale at which organisms perceive the environment (Ritchie 1998, 2002).

Proceeding with the assumption that sampling scale (w) $\propto$ body length (L) and that L is proportional to body mass (M) as $L \propto M^{1/3}$, the models presented in the previous section of this chapter make three qualitative predictions that can be tested with existing experimental and empirical data: (1) minimum cluster sizes of resources (G^*) accepted by consumers should increase with body mass according to $G^* \propto M^{Q/3}$ (eqn. (3.5)), (2) consumption rates of resources C^* should also increase in proportion to $M^{Q/3}$ (eqn. (3.6)), and (3) the minimum supply rate of resources for persistence or resource availability at population equilibrium does not obey a scaling law. As I argued earlier in the chapter, Q^* and q^* change non-linearly and very slowly with increasing w, such that they would appear to be virtually invariant with w and therefore with mass. In the case of G^* and C^*, the scaling exponent should not be a universal constant but should instead reflect the abundance and distribution of the resources in space.

The response of G^*, q^*, and C^* to consumer body size can be tested with data. First, I tested the quantitative prediction that minimum cluster size would increase as $M^{Q/3}$. I compiled data from several literature sources that measured minimum food item size (a surrogate for resource cluster size) and grouped species into two trophic groups: herbivores (insects and mammals that eat plant foliage) and granivores (mammals and birds that eat all or mostly seeds). A priori we might expect that plant foliage is supplied at a higher abundance in a landscape than are seeds and hence we would expect that Q for plant foliage would be higher than that for seeds. If so, then the scaling exponent for the relationship between bite length and mass for herbivores should scale as M^Q and with a higher exponent than the relationship between seed diameter and mass for granivores. Furthermore, the exponents should be less than 2/3 if we assume that resources are fractal (non-random and clustered) and distributed on a two-dimensional landscape ($D = 2$). I found that bite length for herbivores had a scaling exponent of approximately 0.55 ± 0.032 (SE) (fig. 3.6A), which is significantly

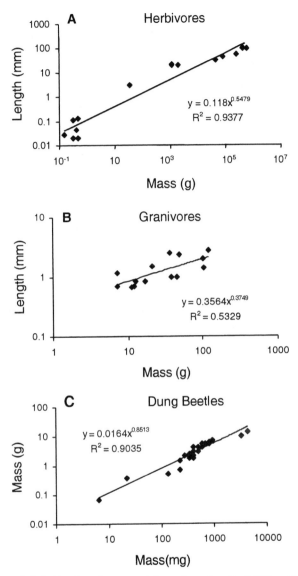

FIGURE 3.6. Scaling laws between minimum item sizes for 3 different functional groups. (A) Minimum item length for herbivores (grasshoppers and mammals) (Belovsky 1984, 1986; Belovsky and Slade 1995), (B) minimum seed diameter for granivores (rodents, Brown 1975; and birds, Grant 1986), and (C) nest-provisioning dung balls for ball-rolling dung beetles (Coleoptera: Scarabeidae). Note the differences in slopes for the three groups.

less than 2/3 and is significantly greater than the scaling exponent for seed diameter, which was only approximately 0.37 ± 0.06 (fig. 3.6B). These exponents should correspond to the fractal dimension of edible plant tissue and seeds, respectively, in the environment.

As a further test, I sampled the dung balls formed by six different ball-rolling dung beetle species (Coleoptera: Scarabeidae) feeding on the dung of white rhinos (*Ceratotherium simum*) in South Africa (fig. 3.6C). These beetle species varied in body mass by four orders of magnitude, with considerable variation within species. Individuals colonize dung patches from which they scour and roll dung balls to provision the nests of their offspring. Relative to the size of even the largest beetle, the dung patch represents a virtually 3-dimensional resource. The mass of balls rolled by individuals therefore would be expected to scale with body mass as M^1. I found that dung ball mass scaled with beetle body length with an exponent of 0.85 ± 0.028 (SE), which is significantly less than 1 and significantly greater than the exponents for minimum bite size for herbivores and minimum seed size for granivores. Because these beetles roll the balls they form as far as 12 m from the dung patch before building a nest, the departure of the observed exponent from a value of 1 may reflect some additional biomechanical constraints associated with rolling balls over and around vegetation or other obstructions.

I also tested a second quantitative prediction of the model: that the resource density at population equilibrium q^* would be approximately invariant with body mass. I used the results of several of my studies of population dynamics and competitive interactions among grasshopper species of different size (Orthoptera: Acrididae) in Minnesota (Ritchie and Tilman 1992, 1993), and Utah (Ritchie 1996). These studies experimentally introduced high densities of grasshoppers into replicate field cages placed over existing vegetation. Grasshoppers typically declined in these cages until they reached a constant density for 14–20 days, after which the experiments were terminated and the green biomass of different plant species inside the cage was measured. By matching observed diets of the different grasshoppers to the final composition and biomass of different plant species in cages, the density of edible plant tissue remaining in each cage (final biomass) was measured. Since these populations were at least at a temporary equilibrium, the final biomass of remaining edible plant tissue represents a

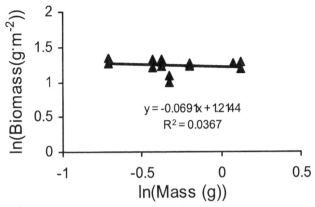

FIGURE 3.7. Green plant biomass remaining in field cages after different grasshopper species, varying by an order of magnitude in body mass, were stocked at high densities and allowed to decline to a constant density. The slope is not significantly different from zero ($P > 0.8$). Data compiled from Belovsky and Slade (1995), Ritchie and Tilman (1992, 1993), and Ritchie (1996).

measure of the R^* of classical consumer-resource models and q^* of the scale-dependent consumer-resource model. Strikingly, there is a strong convergence of the final edible plant biomass for species of different body size. When plotted as a potential scaling law in figure 3.7, the final biomass appears to be nearly invariant with grasshopper body mass, as the slope (-0.062 ± 0.045) is not significantly different from 0 ($P > 0.45$). However, as the scale-dependent consumer-resource model predicts (eqn. (3.10)), there is a slight but insignificant trend towards lower final biomass for larger species.

These results support the predictions of the scale-dependent consumer-resource model for G^* and Q^*. Minimum bite length, seed diameter and dung ball mass all showed relatively tight scaling relationships with consumer body mass. However, there is no apparent universal scaling exponent for G^*, which would be expected if it were determined by physiological constraints, such as metabolic demand (Gross et al. 1993; Brown 1995). Instead exponents vary for different trophic groups in ways that are consistent with the expected density and distribution of their respective resources. These results arise directly from heterogeneity, as the scale-dependence of G^* and C^* disappears if resources are randomly or uniformly distributed such that $Q = 2$. This result suggests that the geometry of resource distribution is a major factor that affects the degree to which different-sized species separate in their resource use. As I will show in subsequent chapters, this geometry has major implications for the coexistence of species and patterns of abundance and species richness.

SUMMARY

1. The dynamics of consumers and their resources can be described as functions of the amount and distribution of resources in space.
2. When resources are distributed in a fractal-like manner, foraging organisms that sample at different spatial scales, of length w, encounter resources at different rates. Larger-scaled consumers are much more likely to find an occupied sampling volume, while smaller-scaled consumers experience higher resource densities per occupied sampling volume.
3. The consumer's sampling scale w determines the resource abundance and distribution at which a species can exist in the environment. This threshold explicitly states the niche boundary of a species because the proportion of landscape occupied by resource when the consumer reaches equilibrium is the same as the supply of resources required for the species to persist or increase when rare.
4. Larger-scaled consumers are expected to avoid the smallest resource cluster sizes at all possible fractal resource densities and distributions, although all types of consumers should become unselective if resources are distributed randomly or uniformly. The difference in the logarithms of minimum cluster size for two consumers that differ in sampling scale will always be proportional to the difference in the logarithms of their sampling scales. This produces a scaling law for minimum resource cluster size for consumers that differ in sampling scale.
5. This scaling law for minimum resource cluster size translates into a similar scaling law for the optimal resource consumption rate.
6. The required resource distribution Q^* and proportion of landscape occupied by resources q^* for persistence (or when the consumer species reaches equilibrium), are analogous to the classic R^*, and are lower for larger-scaled consumers if resource requirements are

constant. However, Q^* is relatively insensitive to sampling scale. Thus, both the resource supply rate and available resource density for persistence and/or equilibrium should decline so weakly with increasing scale as to appear scale-invariant.

7. Minimum food item size (a surrogate for G^*) scales with consumer body size in a manner consistent with the qualitative predictions of the scale-dependent model developed in this chapter. Exponents varied for different trophic groups, suggesting that there is no evidence of a universal scaling exponent for G^*. Likewise, equilibrium edible plant biomass (a surrogate for q^*) in field cage experiments with grasshoppers did not vary significantly with grasshopper body size, as predicted by the scale-dependent consumer-resource model.

8. The geometry of resource distribution is a major factor that affects the degree to which different-sized species separate in their resource use. The scale-dependent consumer-resource model based on this geometry builds a framework for exploring how species with different sampling scales (and thus different body sizes) acquire resources, which ultimately determines community structure and species richness in spatially complex landscapes.

CHAPTER FOUR

Food, Resources, and Scale-Dependent Niches

Thus far I have considered resources as if they were consumed directly, such as when phytoplankton take up dissolved nitrate or phosphate from a water column or when plants harvest photons of light. However, many limiting resources, such as proteins and carbohydrates for animals, are not consumed directly. Instead, they are "packaged" in other material that, for lack of a better term, can be called *food* (Ritchie and Olff 1999) Organisms therefore must consume "food" in order to obtain limiting resources. For example, carnivores eat other animals and herbivores eat plant tissue, seeds, sap, fruits, etc. in order to consume protein. Perhaps less obviously, plants uptake limiting soil nutrients dissolved in water (Lambers et al. 1998), so water could be considered "food" in this context. Microbes and other single-celled organisms may be limited by the energy from electrons contained in the bonds of complex molecules (Battley 1987; Yoshiyama and Klausmeier 2008) or by actual nutrients such as ammonium (Brown and Rose 1969). If so, then the packaging of resources introduces a new source of heterogeneity that influences a consumer's resource intake.

In this chapter, I consider this new source of heterogeneity and its implications for differences in the use of resources by species with different sampling scales, w. The packaging of resources in otherwise nonlimiting material induces a trade-off between maximizing food intake and maximizing resource concentration within food in order to maximize total resource intake. This trade-off yields the qualitative prediction that larger-scaled species must sample volumes with bigger clusters of food while smaller species must sample food clusters with greater resource concentrations. To test these ideas, I then explore patterns in

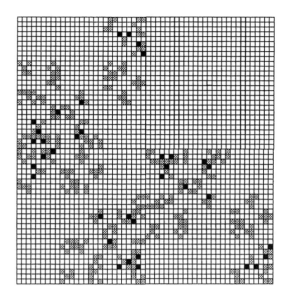

☒ Food: $F = 1.46$, $m = 0.171$
■ Resources: $Q = 0.63$, $q = 0.012$
   Mean resource concentration within food = $q/m = 0.07$

FIGURE 4.1. Hypothetical fractal distributions of a food material (gray + black pixels) and a resource within food (black pixels). Since food fills space faster with increasing scale of observation than does resource-filled food, $F > Q$ and $m > q$. These properties are general whenever resources are "nested" inside food, that is, when the forager cannot use resources outside food.

food consumption by species of different body size, which presumably differ in their sampling scale.

RESOURCE PACKAGING AND THRESHOLDS OF FOOD PATCH SIZE AND RESOURCE CONCENTRATION

If the limiting resource can only be consumed if it is packaged in a food that the consumer can eat, then the resource distribution obviously is constrained by the distribution of food (fig. 4.1). Suppose food has a fractal dimension F and fills a proportion m of the landscape. The resource distribution is necessarily *nested* inside the food distribution, since any resource material that occurs outside food cannot be consumed and is therefore not available. For example, herbivores can obtain nitrogen, a commonly limiting nutrient, packaged in plant tissues, but are of course unable to consume atmospheric nitrogen or dissolved mineral nitrogen in the soil. These forms of nitrogen therefore are not resources; only the nitrogen in plant tissue is a resource because herbivores can only process nitrogen bound in organic chemicals in plants. If resources are distributed with fractal dimension Q and occupy a proportion q of the landscape (as in chapter 3) the "nestedness" of resources implies that *the dimension Q of resources must be less than the dimension of food, F.* Mathematically, this must be true because resources within food fill space less quickly than food as they are observed at larger scales. Resources that occur outside food are considered unavailable so $Q < F$.

If resources are packaged in "food", then the resource consumption rate is a function of the distribution of both food and resources, $C(q,Q,f,F,w)$. Using the logic applied in chapter 3, food will be encountered in any given sampling unit with a certain probability p_1 and resources will be encountered within a sampling volume, given that it contains food, with a probability p_2. However, in this case the probabilities of encountering a particular food cluster of size P or resource concentration R depend on each other. For example, within sampling units occupied by both food and resources, low resource concentrations are more likely in larger food clusters, while high resource concentrations are more likely in smaller food clusters. A given cluster size, $G,$ of resources will occur at lower concentration within food if cluster size P is larger. Conversely, resource concentration is more likely to be high in smaller food clusters. Thus, P occurs with a probability $f(P|R)$ conditional on the choice of $R,$ and resource concentration R will occur in a sampling unit of food cluster size P with conditional probability $f(R|P)$. Modifying the functional response equation for resource intake from chapter 3 to accommodate food cluster size and resource concentration, resource consumption rate,

$$C(q,Q,m,F,w) = vp_F \frac{\displaystyle\int_{P\min}^{P\max}\int_{R\min}^{R\max} Pf(P|R)Rf(R|P)dPdR}{1 + hvp_F \displaystyle\int_{P\min}^{P\max}\int_{R\min}^{R\max} f(P|R)f(R|P)dPdR}. \quad (4.1)$$

Resource encountered is the product of food cluster size P, the conditional probability of that food cluster size given R, $f(P|R)$, the resource concentration R of that cluster, and the conditional probability of different resource concentrations given a certain food cluster size, $f(R|P)$. This product is summed over all food clusters of different sizes P and resource concentrations R. Likewise, the fraction of time spent handling food clusters is the sum of the joint conditional probabilities of encountering different food cluster sizes and resource concentrations used by the consumer.

A fractal-like distribution of food and resources within food constrain these probabilities, as they did for resources in chapter 3. Resource consumption $C(q,Q,m,F,w)$ is determined by the movement of a consumer through a landscape with a fractal-like distribution of resources and its encounter with "local" resource clusters of different densities. A consumer will sample v sampling units in the time period dt over which consumer and resource densities change. In each sampling unit, the consumer will encounter food, in any cluster size and resource concentration, in a sampling volume with probability p_F. This probability is determined as $1 - p_O$, where p_O is the probability that all the w^D cells in a sampling volume are empty. Individual cells are occupied with a probability m or the proportion of the landscape occupied by food, and therefore are empty with probability $1 - m$. By analogy from the derivation of p_E in chapter 3, that is, after substituting F for Q in equation (3.4),

$$p_F = 1 - \left(1 - x^{F-D}\right)^{w^D}. \tag{4.2}$$

For the majority of biologically realistic values of x and F, p_F very quickly saturates to a value of 1 with increasing w. Therefore, for simplicity here I will assume that $p_F = 1$, but I explore the consequences if $p_F \ll 1$ in chapter 7.

The consumption rate function can be further specified by recognizing the constraints of fractal-like food and resource distributions on the conditional probabilities $f(P|R)$ and $f(R|P)$. Bayes' Theorem from basic probability theory gives,

$$f(R|P) = f(P|R)f(P)/f(R).$$

By Korcak's rule (chapter 2) (Hastings and Sugihara 1993), the frequencies of food and resource cluster sizes of exactly size P and G respectively are

$$f(P) = c_P P^{-F/D}, \tag{4.3}$$

and

$$f(G) = c_G G^{-Q/D}, \tag{4.4}$$

where c_P and c_G are calibration constants reflecting the resolution at which sampling scales are discriminated (see chapter 2, box 2.1, eqn. (2.1.1)). Because resources are defined to be nested inside food, a given resource cluster size G requires food cluster size P to at least be of magnitude G. Thus, the probability of a food cluster of exactly size P ($c_P P^{-F/D}$) is drawn from the set of food cluster sizes greater than G, which by Korcak's rule is the frequency $G^{-F/D}$. By definition $R = G/P$, so it follows that $G = PR$. Applying these identities and equation (4.4) yields

$$f(P|R) = c_P P^{-F/D}/G^{-F/D} = c_P R^{F/D}$$

$$f(R|P) = f(P|R)f(P)/f(R) = (c_P/c_G)R^{(F + Q)/D}P^{(Q-F)/D}. \tag{4.5}$$

In chapter 3, I showed that a restricted range of resource cluster sizes should be consumed because including smaller clusters imposes an opportunity cost of the time required to handle small clusters. As I argued above, larger resource clusters tend to have lower resource concentrations because food density increases faster with scale than resource density ($F > Q$). On the other hand, consuming smaller food clusters provides a lower rate of total resource consumption per unit time even if these clusters have more concentrated resources. This trade-off between the contribution of food cluster size and resource concentration to total resource intake predicts that consumers will select a range of food clusters from some intermediate minimum cluster size to the largest possible cluster size and resource concentrations within these clusters from some intermediate concentration to the highest possible. These ranges are defined by a threshold minimum food cluster size P^* relative to a maximum ($P_{max} = w^F$) and resource concentration R^* relative to a maximum ($R_{max} = w^{Q-F}$). As food clusters of decreasing size and/or resource concentration from

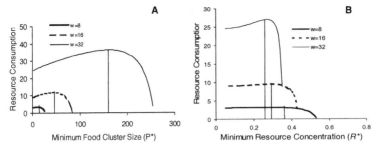

FIGURE 4.2. Changes in resource consumption rate for consumers of different sampling scale w choosing (A) different minimum cluster sizes P and (B) minimum resource concentrations R (note logarithmic scale for P). The shapes of these curves vary so that species of different sampling scale, small (thick curve), medium (dashed curve), and large (thin curve), have different food cluster minimum sizes P^* and resource concentrations R^* that maximize their resource intake. P^* increases with sampling scale, while R^* decreases with sampling scale.

the maximum possible are added to the diet, resource consumption rate increases from zero to a maximum and then declines again to zero (fig. 4.2).

To find these thresholds P^* and R^*, we can substitute our relationships p_1, p_2, $f(P)$, and $f(G)$ (eqn. (4.2)–(4.4)) in equation (4.1), and let the product of the constants $c_P c_G = c$ yield:

$$C(q,Q,m,F,w) = vcp_F \frac{\displaystyle\int_{P^*}^{P\max} \int_{R^*}^{R\max} P^{1+a} R^{1+b} dP dR}{1 + hvcp_F \displaystyle\int_{P^*}^{P\max} \int_{R^*}^{R\max} P^a R^b dP dR}, \qquad (4.6)$$

where $a = (Q - F)/D$ and $b = (2F + Q)/D$ and are thus sums of the Korcak exponents for food and resources (box 4.1).

This equation, analyzed in box 4.1, predicts an optimal minimum food cluster size P^* and resource concentration R^* for consumers with different sampling scales (fig. 4.3):

$$P^* = \theta w^F$$

$$R^* = \sigma w^{Q-F}, \qquad (4.7)$$

BOX 4.1

SELECTION OF DIFFERENT FOOD CLUSTERS BY CONSUMERS

We begin with the integral equation for resource consumption rate as a function of the encounter with food and resource clusters of different size, ranging from some selected minimum food cluster size $P*$ and resource concentration $R*$:

$$C(q,Q,m,F,w) = vc \frac{\int_{P*}^{P\max} \int_{R*}^{R\max} P^{1+a} R^{1+b} dPdR}{1 + hvc \int_{P*}^{P\max} \int_{R*}^{R\max} P^a R^b dPdR}, \qquad (4.1.1)$$

where $a = (Q - F)/D$ and $b = (2F + Q)/D$. Evaluating the double integral yields

$$C(q,Q,m,F,w) = \frac{vc\left[\left(R_{\max}^{2+b} - R^{*2+b}\right)\left(P_{\max}^{2+a} - P^{*2+a}\right)\right]}{1 + hvc \dfrac{\left[\left(R_{\max}^{1+b} - R^{*1+b}\right)\left(P_{\max}^{1+a} - P^{*1+a}\right)\right]}{\phi}}, \qquad (4.1.2)$$

where $\tau = (2 + a)(2 + b)$ and $\phi = (1 + a)(1 + b)$.

To find the optimal minimum threshold $P*$, we take the derivative of the function C with respect to $P*$ and set it equal to zero to yield the following polynomial in $P*$:

$$0 = h(1 + a)vc(X/\phi)(P_{\max}^{2+a} - P^{*2+a}) - P*[(2 + a)$$
$$+ h(2 + a)A(X/\phi)(P_{\max}^{1+a} - P^{*1+a})], \qquad (4.1.3)$$

where $X = R_{\max}^{1+a} - R^{*1+a}$. The function C behaves as a hyperbolic (see fig. 4.2), so this polynomial, for the range of possible values of a and b ($0< F <3$, $0< Q <3$, $F \geq Q$) yields only one reasonable solution in the positive quadrant. This solution is determined primarily by the lower order terms, where $P*$ is small. If we ignore the higher order terms in $P*$, we obtain

$$P* = \frac{vchX(1+a)P_{\max}^{2+a}}{(2+a)[vchXP_{\max}^{1+a} - \phi]}. \qquad (4.1.4)$$

(*Box 4.1 continued*)

The maximum possible value of $\phi = 4$, and the exponent $(1 + a)$ is always positive. As discussed in the text, $P_{max} = w^F$ and will be much larger than 4 in almost any biologically interesting case. Likewise v will be of an order $>10^4$ for most biologically feasible cases where consumption is sufficient to meet resource requirements. Thus the equation for P^* simplifies to the scaling relationship:

$$P^* = \theta w^F, \tag{4.1.5}$$

where $\theta = (1 + a)/(2 + a)$, and will be a fraction for all values of a.

An analogous analysis can be done to find the optimal minimum resource concentration R^* from equation (4.1.2) by finding the derivative with respect to R^*, setting it equal to zero, and ignoring the higher order terms in the polynomial to solve for

$$R^* = \frac{vchY(1+b)R_{max}^{2+a}}{(2+b)[vchYR_{max}^{1+a} - \phi]}, \tag{4.1.6}$$

where Y in this case is $P_{max}^{1+a} - P^{*1+a}$. While R_{max} is defined to be between 0 and 1, the factor $vchY$ is likely to be orders of magnitude greater than maximum $\phi = 4$ for most choices of w. If so, the equation for R^* also simplifies to a scaling relationship,

$$R^* = \sigma w^{Q-F}, \tag{4.1.7}$$

where $\sigma = (1 + b)/(2 + b)$.

---

where θ and σ are related functions of Q and F, that are constant with respect to decisions of the consumer about what food and resource clusters to include in the diet.

$$\theta = \frac{1+[(Q-F)/D]}{2+[(Q-F)/D]}, \quad \text{and}$$

$$\sigma = \frac{1+[(2F+Q)/D]}{2+[(2F+Q)/D]}. \tag{4.8}$$

This result has several important implications. First, the thresholds P^* and R^* are scale-dependent, that is, they both change with the sampling scale of the consumer. More specifically, P^* increases with

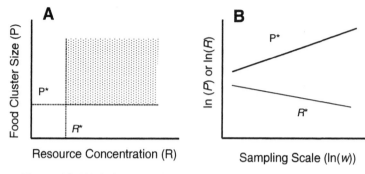

FIGURE 4.3. (A) Optimal selection of minimum food cluster size $P^* = \theta w^F$ and minimum resource concentration $R^* = \sigma w^{(Q-F)}$ that maximize a consumer's rate of resource consumption. These minima define a potential range of acceptable food clusters of sufficient size and resource concentration for a consumer (shaded area). (B) These thresholds follow scaling relationships where the slope for P^* is positive (F) and the slope for R^* is negative ($Q - F$) as indicated by the solutions for P^* and R^*.

sampling scale, implying that larger-scaled consumers should be more selective foragers and choose larger food clusters than smaller consumers (fig. 4.3). Larger-scaled consumers can be more selective because they sample a greater total volume per unit time and thus experience greater opportunity costs from including smaller food clusters in their diet (fig. 4.4). In contrast, R^* decreases with increasing sampling scale (fig. 4.3) because smaller-scaled consumers need to acquire a bigger payoff per unit volume of food, since they sample less total volume and less total food per unit time than larger-scaled consumers. This is true despite the fact that they encounter food and resources at higher *densities*. These two results imply that larger-scaled consumers will use larger food clusters with lower resource concentrations than smaller-scaled consumers. Smaller-scaled consumers will use smaller food clusters that are higher in resource concentration and avoided by larger-scaled consumers. Although P_{max} for larger-scaled consumers is w^F, smaller-scaled consumers likely subdivide such large food clusters into multiple smaller clusters, and use those with higher concentrations of resources and ignore the others (fig. 4.4). Larger-scaled consumers must instead consume the entire food cluster and thus experience a diluted resource concentration. For the same reason, the maximum resource concentration (w^{Q-F}) encountered by larger-scaled

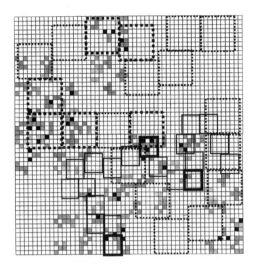

   ▣ Food: $F = 1.46$, $m = 0.171$
   ■ Resources: $Q = 0.63$, $q = 0.012$
     Mean resource concentration within food = $q/m = 0.07$

FIGURE 4.4. Hypothetical encounter of food and resources in the hypothetical landscape of fig. 4.1 for two consumer species, each searching $v = 20$
sampling volumes in a random walk sequence; one with a sampling scale
$w = 4$ (lined squares) and one with $w = 8$ (dashed squares). Sampling volumes that contain food clusters above the optimal thresholds for size and resource concentration for each consumer species are indicated with bold
edges.

consumers is always less than that detected by smaller-scaled consumers (since $Q \leq F$, by definition).

The approximate solutions for P^* and R^* also depend on food and
resource abundance and spatial distribution, as reflected in the coefficients θ and σ and the scaling exponents F and $Q - F$. However, they
do not depend on food or resource demands that might differ between
habitats or seasons. Simulations for a hypothetical forager with $w = 25$
(fig. 4.5) reveal that P^* is sensitive virtually only to food density, as
reflected by its increase with increasing F. Increasing the resource
density and/or dispersion, as reflected by Q, causes very little change
in P^*, mostly because Q does not appear in the scaling exponent.
These responses suggest that the food cluster size selected by foragers
is relatively insensitive to resource concentration and increases

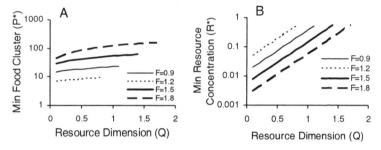

FIGURE 4.5. Sensitivity of (A) minimum food cluster size P^* and (B) minimum resource concentration R^* to changes in food and resource density and distribution. Note that because resources are nested within food and therefore $Q \leq F$, the range of Q values explored increases with increasing F.

dramatically as food density increases. In contrast, R^* increases with increasing Q and decreasing F, as might be expected because both occur in the scaling exponent. Consumers are therefore more selective when resource density is higher or more dispersed and less selective when, for a given resource density, food density is higher or more dispersed. When abundant food leads to lower resource concentrations, as can happen, for example, with increasing biomass of grasses (Olff et al. 2002), consumers can compensate by including food clusters with lower resource concentrations in their diet.

Although less important than the exponents at influencing P^* and R^* values, the coefficients θ and σ reflect a potential scale-invariant selectivity in resource consumption. Crude indices of selectivity are the ratios P^*/P_{max} and R^*/R_{max}, since lower values of these ratios reflect an acceptance of smaller minimum food clusters and lower resource concentrations, respectively, and therefore a wider range of food cluster sizes and resource concentrations, regardless of the consumer's sampling scale. Substituting equation (4.7) and the identities $P_{max} = w^F$; $R_{max} = w^{Q-F}$ leaves $P^*/P_{max} = \theta$; $R^*/R_{max} = \sigma$. A simple inspection of the derivative of θ with respect to Q and F reveals that increasing resource density and/or dispersion (Q), so that food clusters are more homogeneously endowed with resources, causes an increase in selectivity. In contrast, increasing food density and dispersion (F), assuming resources are held constant, causes consumers to be less selective and include smaller food clusters in the diet. Why should this happen?

Increasing F implies that there is simply more food in which a given resource density is distributed, causing mean resource concentration to decline. Recall from our derivation of the conditional probability of P given R that higher resource concentrations are more likely in smaller food clusters. Therefore, including smaller food clusters in the diet is likely to increase the resource concentration per food cluster.

These predicted responses of consumers to a heterogeneous distribution of packaged resources, that is, of food and of resources within food, are consistent with the predictions and experimental results of optimal foraging theory (see Stephens and Krebs 1986; Belovsky 1997). However, because this heterogeneity is assumed to be fractal, the probability distribution of food items or types (interpreted as clusters in this context) can be integrated, which yields a general analytical solution (eqn (4.7)).

NICHE BOUNDARIES FOR DIFFERENT CONSUMERS

The explicit solution to the optimal foraging problem makes it possible to apply the results of resource selection to the determination of niche boundaries for consumers. As I will show, consumers with different sampling scales have access to two sets of resources, one overlapping with consumers of different scales and one that each species can exclusively use. The set of food and resources that can be used exclusively represents an *exclusive niche* for that consumer. Exclusive niches result because the thresholds $P*$ and $R*$ scale, that is change with w, and always rank in the same way for consumers with different foraging scales. Regardless of the minimum supply rates of food or equilibrium food densities, $P*$ is always larger and $R*$ is always smaller for larger-scaled consumers. This means that larger-scaled consumers can use larger, less resource-rich clusters of food than smaller consumers, which can use smaller but more resource-rich food clusters.

This trade-off is seen most clearly from the fact that $P*$ and $R*$ scale with each other. For a given consumer's sampling scale, w, $P*$, and $R*$ are related by w, such that solving for w in the equation for $R*$ and substituting for w into the equation for $P*$ yields

$$P* = \theta\sigma^{F/(F-Q)}(R*)^{-F/(F-Q)}. \tag{4.9}$$

This scaling relationship represents a *lower* bound of food cluster sizes and resource concentrations that, for a given environment (Q and F), is invariant for consumers of different sampling scales. For any community of consumers of different sampling scale, the density and dispersion of food and resources define the foraging niches for all species. These niches are illustrated in figure 4.3 as horizontal (P^*) and vertical (R^*) lines corresponding to equations (4.7).

In a finite foraging environment of extent x, the actual maximum available food cluster size P_{max}, depends on the distribution of food which may be much smaller than that predicted on the basis of fractal geometry (w^F). We can use our equation for the frequency of food clusters of different size based on Korcak's rule to find the actual maximum food cluster size, and thus P^*_{max} in a landscape of extent x. The total amount of food in a landscape is x^F and, the rarest cluster size P_{max} will occur with a frequency P_{max}/x^F on the landscape. The frequency of exactly cluster size P_{max} occurring among all other cluster sizes is $f(P) = c_P P_{max}^{-b}$ where b is the Korcak exponent (F/D). The two frequencies must be equal, so setting them equal and solving for P_{max}, which is also the maximum P^* yields

$$P^*_{max} = c_P^{1/(1 + F/D)} x^{F/(1 + F/D)}, \qquad (4.10)$$

that does not depend on the consumer's sampling scale. This explicitly predicts that *the largest cluster size in a fractal-like distribution increases with the scale of observation*, or interpreted another way, *with the size of the landscape* in which the consumer is searching for resources (fig. 4.6).

What does this mean? Smaller consumers with higher R^* cannot encounter food cluster sizes larger than w^F, since placement of the minimum number of sampling volumes over the distribution of food cannot cover more than that amount of food (see chapter 2). However, such small-scaled consumers can subdivide larger food clusters into subclusters of size w^F, provided these subclusters have sufficiently high resource concentration. Small landscapes, however, contain only a limited number of large clusters, and the largest possible cluster size w^F for a given consumer may not occur with a probability high enough to be found in a small landscape.

Two other constraints on the possible niche space of consumers in such fractal-like environments are the minimum and maximum R^*.

$x = 31, P_{max} = 18$ $x = 54, P_{max} = 25$

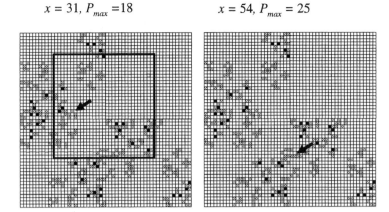

☒   Food
■   Resources

FIGURE 4.6. Effect of increasing the extent of landscape observed, x, on the maximum food cluster size. The hypothetical landscape (same as in fig. 4.1) is "observed" at 2 different extents x, revealing different maximum food cluster sizes P_{max}, indicated by arrows.

A minimum possible R^* corresponds to the R^* where the trade-off constraint intersects P_{max} for the environment (fig. 4.7A)

$$R^*_{min} = \sigma\theta^{1-Q/F}(c_P x^F)^{(Q-F)/(1+F/D)}. \qquad (4.11)$$

A minimum geometrically possible P^* also occurs, with a corresponding maximum R^*. This is defined by the R^* at the smallest possible food cluster size P^* (P^*_{min}) that still allows one cell to be occupied by resource, i.e., $R^*_{min} = 1/P^*_{min}$. Setting $R^* = 1/P^*$ in equation (4.9), solving for P^*_{min} and then for R^*_{max} yields

$$P^*_{min} = \sigma^{-F/Q}\theta^{1-F/Q} \qquad \text{and} \qquad (4.12)$$

$$R^*_{max} = \sigma^{F/Q}\theta^{(F/Q)-1}. \qquad (4.13)$$

The general trade-off constraint of P^* vs. R^* (eqn.(4.9)) and minimum and maximum R^* and P^* generate the set of potential feeding niches for a set of consumers of different sampling scales (fig. 4.7A). This set exhibits a strong negative correlation between P and R, in response to the inevitable trade-off, dictated by a fractal-like distribution

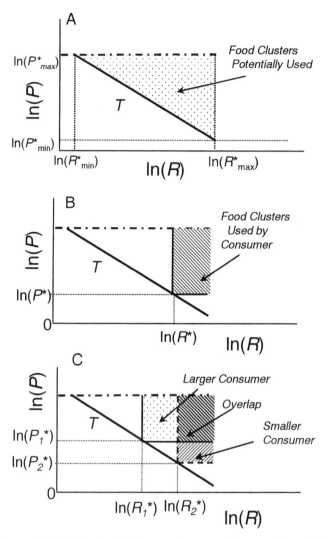

FIGURE 4.7. (A) Hypothetical diagram of resource niche space for multiple consumers with different sampling scales. The line T indicates the combinations of minimum P^* and R^* for consumers of different sampling scale w. P_{max} (alternately dashed lines) is the maximum food cluster size found in the environment. The intersection of T with P^*_{max} defines a minimum acceptable R^* (R^*_{min}), and R^*_{max} occurs at P^*_{min}, the smallest possible food cluster size that can have resource concentration R^* and still contain one cell occupied by resource. (B) Example P^* and R^* that defines a region of potential food clusters of varying size and resource concentration (niche) that will be used

of food and resources, between encountering larger food clusters but lower resource concentrations as sampling scale increases. Although it is possible that large food clusters may have high resource concentrations (upper right-hand corner of niche space), such combinations are extremely unlikely because the joint probability of two rare events, large food cluster and a large resource cluster, is very low.

The set of food clusters actually used by a consumer of a particular sampling scale can be superimposed on this potential niche space (fig. 4.7B). P^* and R^* will intersect the trade-off constraint at a unique point; extending P^* and R^* as horizontal and vertical lines, respectively, to the R_{max} and P_{max} constraints creates a niche boundary. These "niche boundaries" graphically define the set of food cluster sizes and resource concentrations that the consumer will use. For consumers that differ in sampling scale (fig. 4.7C), the minimum P^* and R^* of the largest-scaled consumers will intersect at the far left end of the constraint (small R^*, large P^*) and the smallest-scaled consumers intersecting to the far right (large R^*, small P^*). Immediately, we can see that consumers overlap in using larger, resource-rich food clusters, but that each consumer will use some food clusters that are not used by the other species. More specifically, the model predicts that the larger-scaled consumer *exclusively* will use larger, more resource-poor clusters that are too low in resource concentration for smaller-scaled consumers to use. Meanwhile the smaller-scaled consumer *exclusively* will use smaller, more resource-rich clusters that are too small for the larger-scaled consumer to use. Note also that consumers farther apart in sampling scale will have a larger set of overlapping food clusters, and each will have larger sets of exclusively used food clusters. It is even possible that two consumers of sufficiently different sampling scale might not overlap at all in their use of food clusters. This can occur because larger food clusters cannot contain enough resources to yield resource concentrations high enough to exceed a smaller consumer's R^*, and food clusters small enough to yield resource

by the consumer. (C) Displays feeding niches for two species that differ in sampling scale w: larger consumer (R_1^*, P_1^*) defined by heavy lines; smaller consumer (R_2^*, P_2^*) defined by dashed lines. Stippled region: exclusive resources for the larger consumer. Right-hatched area: exclusive resources for the smaller consumer. Heavy left-hatched region: overlapping resource use.

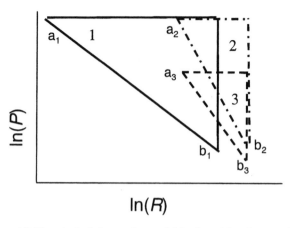

FIGURE 4.8. Hypothetical shapes of potential feeding niches for sets of consumers with different sampling scales, as a function of food abundance and resource concentration (see fig. 4.6). (1) Solid lines: abundant food with low mean resource concentration (High F, Low Q). (2) Alternately dashed lines, abundant food with high mean resource concentration (High F, High Q) (3) Dashed lines, scarce food with high mean resource concentration (Low F, Low Q, $F - Q$ equal to case 2). Lowercase letters, with subscripts for each scenario, refer to critical points that define these shapes: a) R^*_{min}, P^*_{max}; b) R^*_{max}, P^*_{min}.

concentrations $> R^*$ for a smaller consumer, may be too small to be used by a larger consumer.

The environment directly influences the range of possible food clusters that different consumers might use. The trade-off constraint (eqn. (4.9)), as well as P^*_{min}, P_{max}, R^*_{max}, and R^*_{min} (eqns. (4.10)–(4.13)), are sensitive to the density and distribution of both food and resources. However, P^*_{max} is sensitive only to food abundance and distribution, because it is invariant with sampling scale and reflects only the probability of encountering a large food cluster imposed by the size of the landscape. A qualitative analysis of the effects of changing food and resource distributions for a given landscape size, reveals three scenarios (fig. 4.8): (1) high F, low Q (abundant food with low resource concentrations); (2) high F, high Q (abundant food with high resource concentrations); and (3) low F, low Q (scarce food with high resource concentrations). These scenarios illustrate how changing food and resource distributions radically alter the potential niche space of consumers and the way in which sets of exclusive resources differ for consumers with different sampling scale (fig. 4.7).

In scenario 1 the slope of the trade-off constraint is relatively shallow (less negative), implying that for a given sampling scale, a consumer will use smaller minimum food cluster sizes and resource concentrations. Nevertheless, maximum R^* is relatively low and therefore P^*_{min} is relatively large. In addition P^*_{max} is relatively high, yielding a low R^*_{min}. The food clusters used by consumers are shifted toward those with relatively low, on average, and variable resource concentration. In contrast, scenario 2 leads to relatively steep slopes of the trade-off constraint, a relatively high maximum R^* and minimum P^*. This is because increasing resource density (and thus mean resource concentration) and/or dispersion (increasing Q) relative to a given food abundance (F constant), causes the trade-off slope to become steeper (more negative). For a given sampling scale, a consumer will use larger minimum food cluster sizes and resource concentrations. However, P^*_{max}, which is a function only of F (see eqn. (4.10)) and thus increases only with food density and dispersion, is unchanged in scenario 2 relative to scenario 1. Thus, increasing F for a constant Q, that is, increasing food abundance at the expense of mean resource concentration, leads to an expansion in the range of possible food resource concentrations that will be used by consumers of different size. Increasing Q, or mean resource concentration has the opposite effect, as this shrinks the range of resource concentrations available.

In scenario 3, I kept mean resource concentration, represented by the difference $F - Q$, the same as in scenario 2, but reduced both F and Q to simulate an environment with scarce food but high mean resource concentration. In this case, the slope in ln-ln space of T, $[-F/(F - Q)]$, is shallower than in Scenario 2 because F is lower. P^*_{max} is reduced because of lower F, implying that the range of possible food cluster sizes is also reduced. R^*_{max} is approximately the same as scenario 2 and higher than in scenario 1, but P^*_{min} is lower, reflecting the effects of overall lower food abundance. However, R^*_{min} is higher than in scenario 1 but essentially equal to that in scenario 2. Thus the scarce food environment of scenario 3 has the smallest potential feeding niche, with low ranges of both food cluster size and resource concentration.

These niche boundaries imply multiple opportunities for potential resource partitioning by consumers that are all limited by the same resource. This is not analogous to having consumers utilize two distinct resources (Tilman 1982) because food, by itself, is not a resource. The geometry of the feeding niche follows a central scaling relationship between P^* and R^*, the exponents and pre-factors of which reflect both the density *and distribution* of food and resources in the environment. This sensitivity means that the scaling relationships are not universal. Even more interestingly, niche boundaries are sensitive to the size of the landscape (or scale of observation!) in which the consumers live. Therefore, the scale-dependent packaged-resource model of the dynamics of consumers and packaged resources provides a framework for explaining variation in population dynamics, community structure, and biodiversity that I will explore in later chapters.

SCALE, BODY SIZE, AND PATTERNS OF FOOD AND RESOURCE SELECTION

The fractal resource foraging model makes several testable qualitative predictions about the foraging behavior and niche structure of consumers of different sampling scales. As in chapter 3, I will use body size (length, mass) as a proxy for sampling scale, while recognizing, as always, that those behavioral or morphological differences among consumers other than size might create differences in sampling scale. Four major testable qualitative predictions emerge from the model. First, selected minimum food cluster size and resource concentrations should follow a scaling relationship with consumer body size, with food cluster size P^* increasing with body size and minimum resource concentration R^* decreasing with body size. Second, because these scaling relationships are opposite in the sign of their scaling exponents, P^* and R^* trade-off for consumers of different size. Third, this trade-off in P^* and R^* generates exclusive use of food clusters of different size and resource concentration for consumers of different size, with larger consumers using larger, resource-poor clusters and smaller consumers using smaller, more resource-rich clusters. Fourth, the overall niche space potentially used by a community of consumers of the same resource shifts in response to the size of the landscape (or scale of observation), the mean resource concentration, and the abundance and distribution of food. Specifically, increasing landscape size and food abundance dramatically increases niche space, while increasing mean resource concentration restricts niche space.

Scaling of Minimum Acceptable Food Cluster Size and Resource Concentration

The fractal resource model predicts specifically that minimum food cluster size should scale positively with body size, just as generic resources did in chapter 3. Assuming once again that, for L = length and M = mass, $w \propto L \propto M^{1/3}$, so we expect $P^* \propto L^F \propto M^{F/3}$. However, minimum resource concentration is predicted to scale negatively with size: $R^* \propto L^{Q-F} \propto M^{(Q-F)/3}$ (fig. 4.3). The idea of exploring both minimum selected food cluster sizes and resource concentrations is relatively new (Belovsky 1986, 1997; Ritchie and Olff 1999; Ritchie 2002), so there are few data that can be explored to test these two scaling laws. Consequently, I will use Belovsky's data for mammalian herbivores and my own unpublished data for a community of detritivorous dung beetles from South Africa to test these predictions quantitatively.

Belovsky (1997) reported minimum available biomass and quality (soluble carbohydrate plus protein fraction, digested *in vitro* in 2g/L pepsin and 0.01M HCl) of different grassland plant foods (grasses vs. forbs) consumed by a range of 8 different mammalian herbivore species during different seasons (summer vs. winter) at the National Bison Range in Montana. These species ranged in body mass from 0.035 to 450 kg (voles to bison). He repeatedly observed animals foraging on different foods in the field and made detailed measurements of bite size and bite density of plants eaten by the different species over a period of three years. As expected from the fractal resource model, minimum available biomass of plants consumed by the different herbivore species in summer scaled with body mass (fig. 4.9A), with different exponents and coefficients for different foods and in different seasons. Likewise, the minimum fraction of digestible biomass scaled negatively with increasing body mass. However its scaling exponent did not change with different foods or seasons, as all values fit on the same line (fig. 4.9B). Power laws fit very strongly to these data ($R^2 > 0.90$,

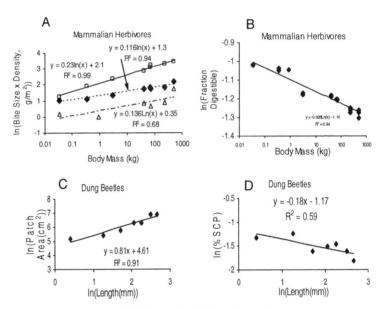

FIGURE 4.9. Scaling of minimum food cluster size and resource concentrations selected by organisms of different size for two communities: (A, B) mammalian herbivores at the National Bison Range (adapted from Belovsky 1997), and (C, D) Hluhluwe-Umfolozi (South Africa) dung beetles. Food cluster size for the mammalian herbivores is bite size x density for: forbs (open squares), grasses in summer (filled diamonds), and grasses in winter (open triangles). For herbivores, resource concentration is the fraction of carbohydrates and proteins digestible in vitro (Belovsky 1997). For dung beetles, food cluster size is dung patch area, and resource concentration is the fraction of soluble carbohydrates and proteins (SCP).

$P < 0.001$), except in the case of winter grasses (open triangles in fig. 4.9A) where sample size was smaller ($R^2 = 0.68$, $P = 0.02$).

I found similar patterns in the patch sizes and resource concentrations of the dung of different ungulate species selected by dung beetles of various families in a South African savanna, the Hluhluwe-Umfolozi Preserve (fig. 4.8C, D). I collected all the dung beetles in or beneath dung piles of 6 different ruminant and non-ruminant species, ranging in size from impala (40 kg) to elephant (3000 kg), in the early dry season of 2002 and sorted these according to species. The dung patches varied in size (area covered) from 50 to 70,000 cm² and in the frequency of different particle sizes (sorted with sieves of 6 different mesh sizes over the range 0.2–20 mm), and the resource (soluble carbohydrate and protein, SCP) concentration of these different particles as determined by digestion of particles in a neutral detergent. Most of the dung beetle species I sampled consume the finer particles of dung and provision nests with dung balls that are later consumed by beetle larvae. Consequently, I used the fraction of SCP in the smallest particle size as a measure of the resource concentration of the dung patch. Since most individual species occurred in less than 50% of dung patches, I sorted species into one of 7 body size classes and then searched for the smallest and most SCP-poor dung patch occupied by beetles in that size class. As predicted by the fractal resource foraging model, these minima scaled with beetle size (body length), with minimum patch area scaling positively (fig. 4.9C) and minimum resource concentration scaling negatively (fig. 4.9D. A power law fit each set of data reasonably well, given the small number of size classes available (patch size: $R^2 = 0.91$, $P = 0.002$; SCP: $R^2 = 0.59$, $P = 0.04$).

A more critical question is whether the scaling exponents (slopes in ln-ln space) are within a range that might be expected from the fractal resource model. Specifically, the exponents for food cluster size should be $F/3$ for mass or F for length. For a 2-dimensional landscape ($D = 2$), the maximum F is 2, so the upper bound to the scaling exponent for food cluster size is 2/3 for mass, 2 for length. For food resource concentration, we would expect the exponent to be $(Q - F)/3$ for body mass and $Q - F$ for body length. As detailed earlier in the chapter, these exponents also imply that minimum food cluster size scales negatively with minimum resource concentration with an exponent $-F/(F - Q)$.

In the case of the grassland herbivores, Belovsky did not measure the spatial distribution of grasses and forbs, but we can use results from Milne et al. (1992), who reported fractal dimensions of 1.6–1.9 for grasses in New Mexico at landscape extents > 1 km. Thus we might expect the exponent for available biomass for the grassland herbivores to lie between 0.53 and 0.63. The exponents for minimum food cluster size for grassland herbivores are much lower than this range (fig. 4.9A) but certainly a fraction significantly greater than zero. In the case of the dung beetles, exponents match up somewhat better. Fresh dung at Hluhluwe Reserve is very patchily distributed on the landscape and occupies, at any given time, less than 0.01% of possible space (Han Olff unpublished data). Assuming an observer scale of resolution of

1 mm and assuming that a given dung beetle searches for dung in at least 1 ha, we can define landscape extent as $x = 10^5$ mm. Using equation (2.5) from chapter 2 and applying it to the case of food distribution, we can estimate $m = x^{F-D}$. Assuming $m = 0.01\%$ or 0.0001 (as an upper bound) and $D = 2$, we can estimate that $F = 1.2$. Consequently the scaling exponent for dung patch area should be less than 1.2, and is exponent = 0.81. Without knowing much more about the actual spatial distributions of food in these empirical examples, it is difficult to be more precise in our expectations for these exponents or to explain why observed exponents for minimum food cluster size were much lower than expected.

If resources (soluble carbohydrates and protein) are more or less uniformly or randomly distributed within food, as might be expected for herbaceous plants and/or their residues in dung, then Q will be very close to but slightly less than F and the exponent for food resource concentration will have a very small negative value. The observed values of −0.028 (($Q - F$)/2, fig. 4.9B) for grassland herbivores and −0.18 for dung beetles ($Q - F$, fig. 4.9D) are consistent with this expectation. Moreover, the scaling exponent should be largely independent of the abundance of food, which would explain why it was similar across seasons for grassland herbivores while the exponent for minimum food cluster size decreased for less abundant food types (forbs) or in seasons when food was less abundant (Fig. 4.9A).

The qualitative prediction that a trade-off between P^* and R^* follows a scaling law also can be tested with these data. Grassland herbivores exhibited different relationships, but all featured strong fits of scaling laws to the data (fig. 4.10, $R^2 > 0.90$ for summer plants, $R^2 = 0.76$ for winter). I argued earlier that Q may be very close to F for resources distributed in herbaceous plant tissue and therefore the scaling exponents for P^* vs. R^* should be much less than −1: $-F/(F - Q)$, with more resource-rich food types (higher Q) having more steeply negative slopes than less abundant types. This is observed for the grassland herbivores (fig. 4.10) for which forbs with higher leaf nutrients had a more steeply negative slope (−8) than did either summer forbs (−3.8), or winter grasses (−5).

The idea of a trade-off between the two minima is not new (Bell 1970; Belovsky 1986; Van Soest 1994), and a foraging model for generalist herbivores that assumes food intake is co-limited by time and

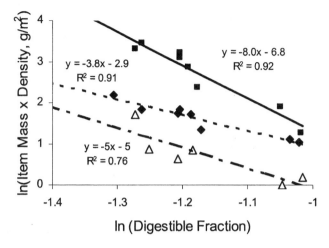

FIGURE 4.10. Minimum item size x density (food cluster size) and resource concentrations (digestible fraction of carbohydrates and proteins) of different plant foods selected by different mammalian herbivore species (0.035 to 500 kg range in body size) at the National Bison Range (adapted from Belovsky 1997). Data are for forbs (filled squares, solid line), grasses in summer (filled diamonds, dashed line), and grasses in winter (open triangles, alternately dashed line).

digestive capacity (Belovsky 1997) explicitly predicts a quantitative relationship for such a trade-off. However, this co-limitation model contrasts with the fractal resource model in not predicting a simple scaling law between minimum food cluster size and minimum resource concentration, although the relationship may approximate a scaling law at high minimum resource concentrations. More specifically, Belovsky fit his co-limitation model to the data in figure 4.10 with non-linear regression and found generally poorer fits (R^2 was 0.10 to 0.20 lower for the same data) than the scaling laws predicted by the fractal resource model. It is impossible at this point to draw much more of a conclusion because the spatial distributions of grasses and forbs were not measured. It is also possible that the two models are driven by similar mechanisms: larger consumers can sample more space or take in a greater volume of food, which requires them to encounter higher food cluster sizes and tolerate lower resource concentrations within food. If so, then the fractal resource model provides an explicit and general extension of the older ideas based on digestive limitation of herbivores

into problems of how consumers respond generally to spatial heterogeneity, even if they are not digestion-limited in their food intake.

The definitive data needed to test these ideas are scarce, largely because the independent distributions of food and resources for many
consumers are often difficult or extremely tedious to measure. Nevertheless, more field data from a wide variety of organisms are needed to
test these models with more precision. In particular, there need to be
measurements of food distributions in space and resource distributions
within food to make quantitative predictions of scaling laws of P^* and
R^* vs. size and vs. each other. Detailed measurements of the use of
different food clusters by consumers of different size are also strongly
needed. In addition, experiments that systematically vary food and resource distributions and measure responses of consumers of different
size are also an important next step.

The relationships between P^* and R^* with body size and their negative relationship with each other implies the existence of exclusive
niches for consumer species of different size. Do these exclusive
niches exist and do they correspond to a trade-off between food quality
and abundance for species of different size? Several studies of pairwise competitive interactions among different herbivore species, including studies of moose and snowshoe hare (Belovsky 1984) and
field experiments with grasshoppers (Orthoptera: Acrididae) (Belovsky
1986, 1997; Chase and Belovsky 1994; Schmitz et al. 1997) demonstrated evidence of exclusive resources. Furthermore, larger species
generally exclusively use larger, lower quality food clusters than
smaller species, and this niche partitioning may be critical for the coexistence of these species. Although minimum food cluster size and
resource concentrations for competing species were measured in some
cases, a description of the sets of food cluster sizes and resource concentrations actually used by different species within a community has
not yet been published.

The data for South African dung beetles, however, explicitly demon-
strate such sets. I plotted the dung patch area vs. the soluble carbohydrate and protein (SCP) content of dung patches used by different size
classes of dung beetles (fig. 4.11). A given size class (e.g., the largest,
fig. 4.11A) uses a clearly delineated set of dung patches. As predicted
by the fractal resource model, both the largest and smallest size classes
select dung patches above particular size and SCP thresholds, that is,

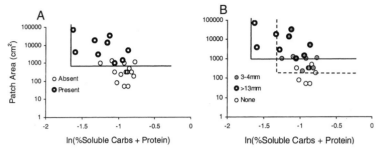

FIGURE 4.11. Distribution of a guild of South African tunneling dung beetles of different size among 26 different dung patches of several ungulate species that varied in size, area, (cm²), and resource concentration (% soluble carbohydrate + protein). (A) Solid line brackets patches occupied by the largest size class (>13mm, open circles with heavy outlines) as opposed to unoccupied patches (open circles with thin outlines). (B) Solid line for the largest dung beetle patches is accompanied by dashed lines that bracket patches occupied by the smallest size class (3–4 mm, filled gray circles). Filled gray circles with heavy outlines are areas occupied by both the largest and smallest classes, while open circles are areas occupied by neither size class. Lines are added to aid the recognition of groupings and are not fit to data.

there are patches avoided by both size classes. The sets of patches used shift from only large patches of virtually all SCP values for the largest beetle size class to only high SCP patches of a wider range of sizes for the smallest size class. Furthermore, the two size classes overlap considerably in their use of large, high-resource concentration patches. These data also show the expected negative correlation among all dung patches between cluster size and resource concentration, as predicted by the fractal resource model. Even these data, however, are too few to analyze statistically, and a much more complete study that samples > 100 dung patches or, more generally dung clusters, is needed to definitively test how species' niches and patch size and resource concentration change with species size.

The limited data available for grassland herbivores and dung beetles suggest that size-dependent resource partitioning, driven by a general trade-off in response to fractal food and resource distributions by consumers of different size, can generate exclusive use of particular sets of food clusters by species of different size. This trade-off depends on the abundance and resource concentration of food, which generate prospective niche space in food cluster size and resource concentration. The data for grassland herbivores suggests that food with higher resource concentration, such as forbs generates a steeper slope in the trade-off between P^* and R^* and potentially a more restricted niche space (fig. 4.8). If so, the partitioning of food clusters according to size and resource concentration by species of different size (or sampling scale!) may be generated by fractal heterogeneity in food and resource distributions. If so, this mechanism of resource partitioning may be an under-appreciated means of coexistence (Ritchie 2002) that may explain organization and diversity in many communities. The limited data correspond sufficiently to the predictions of the fractal resource model that it seems fruitful to take the next step and apply these niche concepts to the structure and diversity of communities in subsequent chapters.

SUMMARY

1. Resources for most species are packaged within other material, which, for lack of a better term, I call "food." Food may frequently have a fractal-like distribution.
2. Consumers may optimize their selection of the range of food cluster sizes and resource concentrations. The fractal resource modeling approach used in chapter 3 can be expanded to include packaged resources. The model predicts that minimum acceptable thresholds of food cluster size (P^*) and resource concentration (R^*) should follow simple scaling laws directly with consumer sampling scale or size but with opposite signs of exponents. The exponents are proportional to the fractal dimensions of food and resources, respectively.
3. The opposite signs of these scaling laws for P^* and R^* imply a trade-off, whereby larger consumers require larger food clusters but can tolerate lower resource concentrations, while smaller consumers require higher resource concentrations but can tolerate smaller food cluster sizes. This trade-off yields into two major qualitative predictions: (1) species of different sampling scale or size will have exclusive use of sets of food clusters with particular size and resource concentration; and (2) the scaling law for P^* and R^*, plus upper bounds on maximum food cluster size set by the scale of ob-

servation or extent of the "landscape" in which species occur, define niche space for a community of different-sized consumer species.

4. Niche space available to a community of species of different sampling scale or size that consumes the same resources changes with the abundance of food and its resource concentration, as well as with the scale of observation or extent of the "landscape" in which species occur. For a given landscape extent, the largest niche space occurs with abundant, high fractal dimension food but a low fractal dimension of resources.

5. Limited empirical data from two independent datasets, grassland herbivores and dung beetles, are fit qualitatively quite well ($R^2 > 0.90$ in most cases) by simple power laws, as predicted by the fractal resource model. Minimum food cluster size scales positively with size, while minimum resource concentration scales negatively with size, each with exponents in the range of values expected from the fractal resource model. However, the exponents for minimum food cluster size were smaller than expected from available data in the literature on plant and dung distributions; this discrepancy indicates the need for more studies of resource use thresholds together with measurements of food and resource distributions. These minima also were negatively related to each other according to a simple scaling law, and had estimated exponents within the range expected from both the fractal resource model and the scaling laws for minimum food cluster size and resource concentration with body size.

6. The simplicity of the solutions of the fractal resource model and the agreement of two independent data sets make it seem fruitful to apply these niche boundary predictions to predicting the structure and diversity of communities, which I address in future chapters.

CHAPTER FIVE

Size Structure in Ecological Guilds

Three fundamental and inter-related characteristics of ecological guilds, or communities of species that use the same resources, are the number of species of different sizes, the abundance of different-sized species (Hutchinson and MacArthur 1959; Morse et al. 1985; Damuth 1991; Blackburn and Gaston 1999; Brown et al. 1993; Brown 1995; Siemann et al. 1996) and the limit to similarity in species traits (Hutchinson 1959; Abrams 1975; Pacala and Tilman 1994; Belovsky 1997; Kinzig et al. 1999). Defined more generally as the size structure of guilds, these two characteristics may reflect the outcome of competitive interactions within guilds, although other mechanisms, such as predation or colonization limitation, may influence the size structure of communities (Etienne and Olff 2004). Because of its classical connection to competitive interactions, limiting similarity has received much less attention, but its importance is still recognized (Pacala and Tilman 1994; Abrams 1996, 1999). As I will show in this chapter, the fractal resource model applied to competition theory also predicts limits to the abundance and similarity in size for different-sized species, based on selection of food clusters of different size and resource concentration by species of different size or sampling scale (see chapter 4). Patterns in body size and abundance for several different organisms, including birds, insects, and mammals, support many of the qualitative predictions of the fractal resource model for limiting similarity, abundance vs. size, log-rank abundance relationships, and abundance-frequency distributions. These results suggest that the size structure of guilds is strongly driven by heterogeneity and scale of observers and organisms.

HYPOTHESES FOR THE SIZE STRUCTURE
OF COMMUNITIES

Patterns in the abundance of different-sized species are well studied, but few mechanisms have been proposed to explain them. Hutchinson and MacArthur (1959) observed that both abundance and diversity for different body size classes declined as a scaling law proportional to L^{-2}, where L is length. They hypothesized that abundance and diversity might be inversely proportional to metabolic energy demand. Damuth (1991) later formalized this hypothesis as the "energy equivalence rule." With a larger dataset than was available to Hutchinson and MacArthur, Damuth observed that the maximum abundance of different-sized mammal species from a continental sample scaled with mass (M) as $M^{-3/4}$. Damuth proposed that resources were equally available, and thus equivalent, for species of different size; and that abundance was simply proportional to available resources divided by some multiple of basal metabolic rate, which scales as mass to the 3/4 power for a variety of different organisms, including plants (Schmidt-Nielsen 1983; Peters 1983; Calder 1984; West et al. 1997, 1999; Enquist et al. 1999). This energy equivalence rule provides a key connection between a theory of the metabolism of organisms and community and ecosystem ecology (Allen et al. 2002; Brown et al. 2004; Gillooly et al. 2005).

Morse et al. (1985) proposed an alternative hypothesis that the abundance of organisms would be proportional to the amount of surface area available in their habitat. They measured surface areas of vegetation and found them to be fractal (see chapter 2), and showed that longer species should experience less surface area than smaller species and so should be less abundant. Their model, based on the observed fractal dimensions, D, of vegetation, predicted abundance to be proportional to $L^{-2.4}$, where L is the body length of the insect. Interestingly, if length is assumed to be proportional to $M^{1/3}$, then this would predict that abundance should scale approximately as $M^{-2.4/3}$ or $M^{-0.8}$, an exponent very close to those observed for mammals by Damuth and for insects by Morse et al. (1985). Note that Morse et al. did not assume any influence of resource or energy demand or any differences in this demand for different body sizes.

These two hypotheses predict essentially a negative exponential distribution of abundance with body size and therefore do not predict the left tail observed in many well-sampled abundance or diversity distributions with size (Brown 1995; Siemann et al. 1996; Gaston et al. 2001). The simplest alternative model that predicts a unimodal distribution is that species in a community are limited by multiple interacting factors that act in a multiplicative fashion. Such multiple multiplicative factors should, averaged over enough species, produce the commonly observed lognormal distribution of abundance (May 1975, 1988). Brown, Marquet, and Taper (1993) and Maurer (1998), however, observed right-skewed distributions of abundance for mammals and birds. They formulated a hypothesis that vertebrate abundance and diversity depend on the available energy of individuals to make offspring, or "reproductive power." Smaller vertebrates are expected to decline in diversity because they cannot consume enough energy to make new offspring due the demands of their high mass-specific metabolic rates. Their diversity and abundance should increase with their ability to harvest resources, which is predicted to increase as $M^{3/4}$. In contrast, larger mammals are limited by their ability to convert resources into offspring, which should be proportional to mass-specific metabolism and decline as $M^{-1/4}$. Because of the big difference in the slopes of these two scaling laws for reproductive power, this hypothesis predicts a unimodal distribution of abundance and species richness that is strongly right-skewed, that is, has a long tail to the right of the mode. It predicts a very different distribution from the lognormal, and its right tail, while predicted to be a scaling law, has a very different exponent, $-1/4$, than the size-abundance distributions predicted by the energy equivalence or the habitat surface area hypotheses. This model also makes no explicit prediction about how reproductive power translates into what it is used to predict, namely abundance or diversity.

Etienne and Olff (2004) predict a unimodal distribution of abundance and diversity with size on the basis of colonization vs. resource limitation of abundance. In their model, the probability of colonizing a particular space is judged to be a scaling law of increasing body size, that is, larger species can move farther and are better colonizers. Once in a habitat, individuals are subject to the energy equivalence rule. Diversity arises from neutral dispersal (Hubbell 2001) by species whose abundance, and thus probability of being a disperser, is set by the energy equivalence rule. Depending on the overall dispersal ability of species and resource concentrations in the environment, this model can generate a wide variety of unimodal size-abundance and size-diversity distributions that can be either left- or right-skewed.

We learn several things from these different models and observed data: (1) there is no universal size-abundance or size-diversity distribution; (2) different interacting mechanisms can produce an array of different distributions; (3) there are no predictions about the effects of explicit species interactions such as competition or predation; (4) models and data have been applied to species of similar taxa (mammals, birds, insects) rather than guilds, or those that use similar resources; (5) we don't know why some observed distributions are unimodal, with various skews, and some are essentially negative exponential distributions. These uncertainties beg for an alternative hypothesis about the body size structure of communities from a model of species interactions. The niche-based approach developed in chapter 4 provides a tool by which we can predict size-abundance and size-diversity distributions that result directly from competition for heterogeneous shared and exclusive resources at different scales of observation. The patterns generated from the model can then be compared with those of other models and with observed data.

COMPETITION FOR SHARED AND EXCLUSIVE RESOURCES

The niche space available for species foraging for heterogeneous, packaged, and fractally distributed resources hints at a potential community structure with a large, but not infinite, number of coexisting species (see chapter 4). The critical step in translating niche space into community structure is to determine limiting similarity. Darwin (1859) first recognized that selective foraging by individuals of a given species on a heterogeneously packaged (different prey sizes) or distributed (different patch sizes) resource can reduce their niche overlap with individuals of another species and aid coexistence (see chapter 1). It wasn't until almost a century later, however, that first Huxley (1942) and then Hutchinson (1959) galvanized community ecology with the simple notion that there must be a limit to how similar species can be in one or more traits. Hutchinson proposed that guilds would feature constant size ratios between pairs of species of adjacent rank in body size, as illustrated by the famous aquatic beetles in the pool at the shrine of Santa Rosalia. Analytical solutions for limiting similarity within the standard consumer-resource model were made by MacArthur (1969) and again by Abrams (1975, 1996, 1999) but the predictions were not related explicitly to body size. A very recent model (Yoshiyama and Klausmeier 2008) incorporates both body size and resource cluster size in the form of bacterial cell sizes and resource molecule sizes; but as long as all species are assumed to consume all forms of the resource, just at different rates, then coexistence of a large number of species on a single limiting resource is still highly unlikely.

A more overlooked aspect of the work in the 1960s and 1970s on competitive overlap (Cody and Diamond 1975; Diamond and Case 1986), and a major outcome of the consequence of species sampling the environment at different spatial scales (see chapter 4), is the existence of exclusive resources. When species do not overlap in their use of different resource types, patches, or clusters, as I have described in this book, this generates *exclusive resources* for each species. Divergence in resource overlap does *not* translate only into different *rates* of resource consumption, as assumed even by Levins (1968) in his derivations of competition coefficients for the Lotka-Volterra competition model. Rather, divergence results in ever-larger sets of exclusive resources for species (Ritchie 2002).

The key question for community structure in the context of a heterogeneous fractal niche space becomes, how many exclusive resources must be available for a species to coexist with other species that are either smaller or larger in sampling scale? The development of optimal foraging theory in the late 1960s and early 1970s (Emlen 1966; MacArthur and Pianka 1966; Schoener 1971; Covich 1972; Pulliam 1974; Charnov 1976) was stimulated in part by the desire to predict the separation in species' diets that might lead to coexistence. The qualitative predictions developed in chapter 4, however, explicitly predict that optimal diet, that is, the range of cluster sizes of food and resources within food that maximize resource intake, should differ between species with different sampling scales. Thus, competitive interactions, coupled with the advantages of maximizing resource intake, may yield a co-evolutionary landscape of selection for differences in size or other morphological characters that allow species to sample at different scales and use different sets of exclusive resources.

The condition for a species' persistence in a community can be expressed as a question of limiting similarity, i.e., how different in sampling scale must species be to have sufficient exclusive resources to persist. To answer this question, we can turn to models of competition for shared and exclusive resources, which were worked out for two species nearly 30 years ago by Schoener (1976, 1978). Theoretical ecologists have largely ignored these models, perhaps because they are strongly non-linear and not very mathematically tractable. However, as I will show, foraging trade-offs faced by species with different sampling scales impose strong constraints on the parameters of these models that make them more amenable to predicting community structure. In any case, they are the only available models with the appropriate mechanism for understanding coexistence in heterogeneous, fractal niche space.

The structure of a fractal niche space, that is, food cluster sizes and resource concentrations characterized by maximum food cluster size P_{max} and minimum resource concentration R_{min}, predicts that each species will compete most strongly with the species most similar to it in sampling scale w. Thus, we can rewrite Schoener's models in the consumer-resource framework presented in chapters 3 and 4 to describe the dynamics of a consumer species i, with n competing species for shared and exclusive resources. The rate of population growth of a particular species i is

$$\mathrm{d}N_i/\mathrm{d}t = q_i N_i \left\{ \left[\sum_{j \neq i}^{n} I_{ij} /(N_i + \beta_{ij} N_j) \right] + E_i /N_i - B_i \right\}, \qquad (5.1)$$

where I_{ij} is the supply rate of resource shared between species i and j, and E_i is the supply rate of species i's exclusive resource. Species i converts resources into new individuals at efficiency q_i and loses resources due to maintenance metabolism and mortality at rate B_i. The ratio β_{ij} is the ratio of resource consumption of shared resources for species j relative to that of species i.

The system of equations for n species implied by equation (5.1) cannot usefully be solved at equilibrium, since the dynamical equations

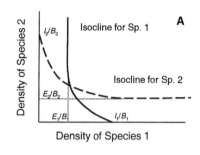

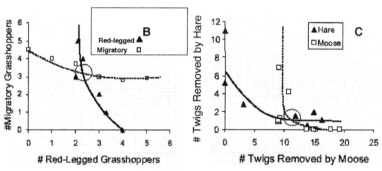

FIGURE 5.1. (A) Hypothetical competitive isoclines for a pair of species competing for shared and exclusive resources. Increasing the density of a competitor (species j) leads to a decline in the steady-state density of a target species i from its equilibrium density when alone (Ii/Bi) to an asymptote (Ei/Bi), the density that can be maintained by that species' exclusive resources Ei alone. Evidence from the field: (B) experimentally determined isoclines and expected equilibrium point for two competing grasshopper species in replicate field cages in Montana grassland (Belovsky 1986; redrawn in Ritchie 2002), and (C) moose and snowshoe hare (Belovsky 1984) on islands near Isle Royale, Michigan (data from Belovsky 1984; drawn by Ritchie 2002), assuming that the density of twigs eaten by snowshoe hare (45° cut) and moose (shredded cut) are proportional to population density.

for each species yield isoclines, or combinations of N_i, N_j at which $dN_i/dt = 0$, that are third-order polynomials with an asymptote at a minimum $N_{i,min} = E_i/B_i$ and an N_i intercept at I_i/B_i, where I_i is the supply rate of all resources available to species i, given its sampling scale (fig. 5.1A). If competition commonly occurs for shared and exclusive resources, then this provides a major new tool for a more general understanding of competition that can be applied to an entire guild. Competitive coexistence is dictated by constraints on the amount of exclusive and shared resources for two species. With exclusive resources, competitive isoclines flatten to an asymptote, implying that an infinite increase in density of the competitor cannot reduce a target species' density below that sustained by its exclusive resources. Such asymptotic isoclines are probably the best evidence for exclusive resources, and have been found in several experiments (Belovsky 1984, 1986; Chase 1996; Schmitz et al. 1997; Ritchie 2002) (figs. 5.1B, C). However, few researchers acknowledge their existence and/or search for them in studies of interspecific competition.

Chapter 4 lays the groundwork for the next step, which is to recognize that species with different sampling scales have different sets of exclusive resources (see fig. 4.7C). In the sections below, I apply the previously developed models of foraging and niche boundaries developed in chapters 3 and 4 for packaged, fractally distributed resources. I first derive the equilibrium abundance of consumers of different sampling scale. I then predict the limiting similarity, or limit to how similar in sampling scale species can be and still persist under competition. With this similarity rule, I then predict the species richness and the size-diversity distribution for a guild.

Community Structure Under Competition for Shared and Exclusive Resources

The equilibrium abundance of all species is virtually impossible to obtain analytically from the general equation for competitive dynamics for n species (eqn. (5.1)). With some reasonable simplifying assumptions, however, one can approximate the equilibrium abundance for a species of a given sampling scale w. The first assumption is that any given species will overlap in resources most and therefore compete most strongly with the species most similar to it in sampling scale. I simplify the general equation to explore the dynamics of a species j, which is larger in sampling scale than a species i but smaller than species k. Because I am exploring competition when species are as close as possible in scale, they share virtually the same set of resources. Thus, I assume $I_{ji} = I_{jk} = I_{ijk}$. The rate of population growth for species j is then

$$dN_j/dt = q_j N_j \{ [I_{ijk}/(N_j + \beta_{ji}N_j + \beta_{jk}N_k)] + E_j/N_j - B_j \}, \qquad (5.2)$$

where I_{ijk} are the resources of species j shared with both the smaller (species i) and larger species (species k), and β_{ji} and β_{jk} are the respective relative consumption rates of species i and k for these shared resources relative to species j. A second assumption is that, since these closest competitors are similar in sampling scale, their equilibrium abundances are likely to be similar, so $N_i^* \cong N_j^* \cong N_k^*$. Thus at equilibrium, where $dN_j/dt = 0$, the equilibrium population density, or abundance, of species j satisfies the following equation:

$$0 = (1/N_j^*)[I_{ijk}/(1 + \beta_{ji} + \beta_{jk}) + E_j] - B_j. \tag{5.3}$$

Solving for N_j^* yields

$$N_j^* = [I_{ijk}/(1 + \beta_{ji} + \beta_{jk}) + E_j]/B_j. \tag{5.4}$$

To develop this further, recall from chapter 4 that the consumption rate for a consumer of a given sampling scale and conditional probabilities of encountering food of a particular cluster size P and resource concentration R is (eqn. (4.6))

$$C(Q,F,D,w) = vcp_F \frac{\displaystyle\int_{P^*}^{P_{max}} \int_{R^*}^{R_{max}} P^{1+a} R^{1+b} dPdR}{1 + hvcp_F \displaystyle\int_{P^*}^{P_{max}} \int_{R^*}^{R_{max}} P^a R^b dPdR}, \tag{5.5}$$

where $a = (Q - F)/D$ and $b = (2F + Q)/D$ are sums of Korcak exponents for food (F/D) and resource (Q/D) distributions. Recall from chapter 4 that the probability of encountering food in a sampling unit of volume w^D is

$$p_F = 1 - \left(1 - x^{F-D}\right)^{w^D}. \tag{5.6}$$

The probability p_F will be nearly 1 in most cases. As in chapters 3 and 4, I assume here that, at competitive equilibrium, food and resource density will be low and that handling time has relatively little impact on resource consumption. Thus,

$$C(q,Q,F,w) = vc \int_{P^*}^{P_{max}} \int_{R^*}^{R_{max}} P^{1+a} R^{1+b} dPdR. \tag{5.7}$$

Because the consumption rates of interest for each of species j's competitors are only for the resources they share with species j, and these shared resources are assumed to be equal, the integral is the same for any pair of species, such that

$$\beta_{ji} = C_i(q,Q,F,w_i)/C_j(q,Q,F,w_j) = (w_i/w_j)^{D-Q}$$

$$\beta_{jk} = C_k(q,Q,F,w_k)/C_j(q,Q,F,w_j) = (w_k/w_j)^{D-Q}. \qquad (5.8)$$

This shows that the relative consumption rates for *shared* resources are a simple ratio of the scaling relationships for the probability of encountering a sampling window occupied by resource. Note that $\beta_{ji} < 1$ since $w_i < w_j$ by definition and likewise $\beta_{jk} > 1$ since $w_k > w_j$. This suggests that competition for shared resources is asymmetric; larger-scaled species have faster encounter rates with the shared set of food clusters, and depending on loss rates, might outcompete smaller-scaled species for shared resources. Note also, as I have shown before, that if the resource is not distributed as a fractal and is random or uniform, $Q = D$ and the system collapses to a model of species competing for completely shared resources and its expected competitive exclusion (MacArthur 1969; Tilman 1982).

Another assumption can be made that further simplifies the system and links the calculation of equilibrium abundance to limiting similarity and species richness. This assumption is that for a given species, the ratio of sampling scales with its nearest-scaled competitors is the same: $w_j/w_i = w_k/w_j = \gamma$. Since the trait that separates species and potentially allows coexistence is the difference in sampling scale, the ratio γ is the critical measure for limiting similarity. Recognizing that $\beta_{ji} = (w_i/w_j)^{D-Q} = (\gamma)^{Q-D}$ and substituting into equation (5.5) yields

$$N_j^* = [I_{ijk}/(1 + \gamma^{Q-D} + \gamma^{D-Q}) + E_j]/B_j. \qquad (5.9)$$

The supply of exclusive resources available to a consumer is that detected by the consumer in time dt, or the time needed for resources to renew once they are consumed. Recall from chapters 3 and 4 that a consumer samples a total volume V, or the instantaneous sample volume w^D multiplied by v samples in time dt: $V = vw^D$. Each sample volume contains *exclusive resource* with probability p_E, or the probability of encountering food clusters that are smaller ($P_j^* < P_k^*$) than those used by species k and lower in resource concentration ($R_j^* < R_i^*$) than those used by species i. This is the double integral (sum) of conditional probabilities we explored in chapter 4, (eqns. (4.5) and (4.6)), evaluated from R_j^* to R_i^* and from P_j^* to P_k^* :

$$p_E = c \int_{P_j^*}^{P_k^*} \int_{R_j^*}^{R_i^*} P^a R^b dPdR, \qquad (5.10)$$

where a and b are the same sums of Korcak exponents for food and resources as we used in chapter 4: $a = (Q - F)/D$; $b = (2F + Q)/D$.

The total exclusive resources in the landscape are just the resources in volume V multiplied by the number of volumes of size V in the landscape that "cover" available resources. The latter is just the available food in the landscape x^F divided by the total sampling volume per individual vw^D. Thus,

$$E_j = vw^D p_F p_E (x^F / vw^D) = cp_F x^F \int_{P_j^*}^{P_k^*} \int_{R_j^*}^{R_i^*} P^a R^b dPdR. \qquad (5.11)$$

Evaluating the integral, substituting equations (4.7) for the respective P_j^*, P_k^*, R_i^*, and R_j^*, recalling that $w_k = \gamma w_j$ and $w_i = w_j/\gamma$, and simplifying yields

$$E_j = \frac{\sigma^{2+b} \theta^{2+a} cp_F x^F}{(1+b)(1+a)} w_j^{F(a-b)+Q(1+b)} \left(\gamma^{F(1+a)} - 1\right)\left(\gamma^{(F-Q)(1+b)} - 1\right). \qquad (5.12)$$

This same approach can estimate the amount of shared resources I_{ijk} as the amount of resources in food clusters larger than the minimum size (P_k^*) for the larger species k and in clusters that are higher in resource concentration than the minimum (R_i^*) used by the smaller species i. The maximum possible food cluster sizes and resource concentrations remain those available in the environment for all species, $P_{max}^* = c_p x^{F/(1+F/D)}$ and $R_{max}^* = \sigma^{F/Q} \theta^{(F/Q)-1}$ (see chapter 4, eqns. (4.11) and (4.13)).

$$I_{ijk} = c_p p_F x^F \int_{P_k^*}^{P_{max}} \int_{R_i^*}^{R_{max}} P^a R^b dRdP \qquad (5.13)$$

Following similar steps as for exclusive resources yields

$$I_{ijk} = cp_F x^F [P_{max}^{(1+a)} - \theta^{(1+a)} (\gamma w)^{F(1+a)}]$$
$$\cdot [R_{max}^{(1+b)} - \sigma^{(1+b)} (\gamma / w)^{(F-Q)(1+b)}]/(1+a)(1+b). \qquad (5.14)$$

These solutions for E_j and I_{ijk} look cumbersome, but they provide a framework for exploring how competition for a single packaged, heterogeneously-distributed resource might structure a community. Next we take these results and explore predicted limiting similarity and then various species abundance relationships.

Limiting Similarity Under Competition for Shared and Exclusive Resources

A species j can persist as long as $N_j^* \geq 1$. Coexistence is guaranteed when all competing species can persist on their respective exclusive resource regardless of the competitive effects of other consumer species. This will be true when the supply rate of a species' exclusive resource in the total sampling volume of an individual E_j^1 is abundant enough to compensate for its resource loss rate, B_i: $E_j^1 > B_i$. We can find E_j^1 by dividing the exclusive resources in the entire landscape E_j by the number of total sampling volumes covering resource-occupied space in the entire landscape (see equation (5.11))

$$E_j^1 = v w^D E_i / x^F$$

$$E_j^1 = \frac{v \sigma^{1+b} \theta^{1+a} c p_F}{(1+b)(1+a)} w_j^{F(a-b)+Q(1+b)+D} \left(\gamma^{F(1+a)} - 1 \right) \left(\gamma^{(F-Q)(1+b)} - 1 \right).$$

$$(5.15)$$

The limiting similarity, γ, that is, the smallest ratio of sampling scales for consumers of adjacent rank in sampling scale, that satisfies this relation cannot be solved for explicitly. However, numerical solutions suggest that γ is usually closer to 1 than to 10 and consequently, for a wide variety of F, Q, and γ values, the polynomial $(\gamma^{F(1+a)} - 1)$ $(\gamma^{(F-Q)(1+b)} - 1)$ is approximately equal to $(\gamma^{F(1+a)} - 1)^2$. With these assumptions the limiting similarity is now solved straightforwardly by substituting for E_j^1 in the inequality $E_j^1 > B_i$ and solving for γ:

$$\gamma(w) > \{ 1 + [(B/c)(1 + a)(1 + b)\theta^{-(1+a)}\sigma^{-(1+b)}v^{-1}$$
$$\cdot w^{F(b-a)-Q(1+b)-D}]^{1/2} \}^{1/(F(1+a))}. \qquad (5.16)$$

At this point we need to consider resource requirements B in more detail before we can elucidate further how limiting similarity should change with food and resource densities and distributions. The well-known allometric relationship between mass and metabolic rate (Peters 1983; Calder 1984) may serve as a good first approximation of B for our purposes. Although the relationship typically refers to units of energy, metabolic rate likely correlates with the demand for other resources such as nitrogen, phosphorus, water, or other potential resources (Peters 1983; Brown et al. 2004). Hence $B = B_0 M^\beta$, where B_0 is a re-normalization constant (essentially metabolic rate for an organism of mass of 1). As we have argued in many of the previous chapters, we might expect mass and sampling scale to be correlated: $M \propto w^3$. Substituting for M leaves: $B = B_1 w^{3\beta}$ where B_1 is a new re-normalization constant that includes a constant for how sampling scale is converted into mass. We can now re-write our equation for limiting similarity as:

$$\gamma(w) > \{ 1 + [(B_1/c)(1 + a)(1 + b)\theta^{-(1+a)}\sigma^{-(1+b)}v^{-1}$$
$$\cdot w^{F(b-a)-Q(1+b)-D+3\beta}]^{1/2} \}^{1/(F(1+a))}. \qquad (5.17)$$

Interestingly, γ is a function of the sampling scale w and thus changes with w and should not, except under very specific conditions, be constant with sampling scale, as predicted by Hutchinson (1959). As expected, it has a minimum of 1, where two species would have identical sampling scales. However, the supply of food and resources determines whether γ increases or decreases with w, yielding a wide variety of different possible relationships (fig. 5.2). Solving for the partial derivative of $\gamma(w)$ with respect to w, it is easy to show that $\gamma(w)$ decreases with increasing w in two-dimensional systems ($D = 2$) and with allometric scaling of resource requirements ($\beta = 3/4$) approximately when $Q > F - 1/2$.

This means that under many biologically realistic conditions, larger-scaled species can be relatively closer together in scale and still coexist (fig. 5.2A) because the ratio γ of foraging scales for two adjacent-scaled

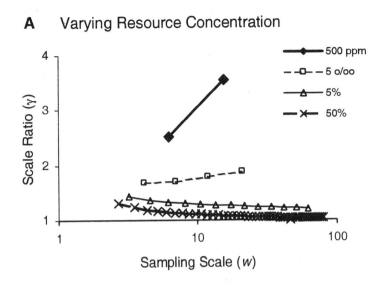

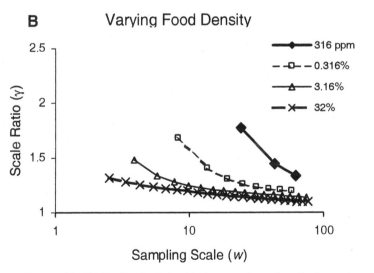

FIGURE 5.2. (A) Predicted relationship between the scale ratio (larger to smaller) of ranked adjacent-scaled species in a community for different mean resource concentrations within food (changing Q independently of F). (B) Predicted relationships between the scale ratio (larger to smaller) of ranked adjacent-scaled species in a community for different mean food densities at fixed resource concentration (changing F independently of Q).

species will be smaller for larger-scaled species. Larger-scaled species can be more tightly packed in a community because their exclusive resources occur at higher density and thus can be partitioned more finely with their nearest competitors. This higher density of exclusive resources for larger-scaled species is evident in equation (5.12) from the exponent of w_j, $F(a - b) + Q(1 + b)$, which is greater than zero approximately whenever $Q > F - 1/2$. Essentially, larger-scaled species have more dense exclusive resources, because with fractal food and resources, larger food clusters with lower resource concentrations occur with higher probability than smaller food clusters with high resource concentration.

Equation (5.17) implies that there are exceptional conditions, the opposite of those shown, that can cause $\gamma(w)$ to *increase* with w (fig. 5.2A). This occurs when food is abundant and is very low in resource concentration ($F - Q > 0.5$). Under such conditions, large food clusters do not differ very much in their resource concentration. Smaller species obtain the high resource concentrations in food clusters that they need by subsampling within food clusters, a strategy that only they can employ. Larger-scaled consumers, rather than escape competition from the next smaller-scaled species by selecting food clusters with lower resource concentration, have to differentiate more in their use of different food cluster sizes. Thus, extremely low food resource concentrations cause larger-scaled species to be further apart in scale than smaller-scaled species.

Under most conditions, limiting similarity $\gamma(w)$ declines rapidly with increasing w to approximately an asymptote of low, $\gamma(w)$ (fig. 5.2). Modifying food density and resource concentrations can change the shape of this relationship. Decreasing food density while holding resource concentration constant, and thus decreasing F while adjusting Q to keep $F - Q$ fixed, increases the overall mean scale ratio among adjacent-scaled species pairs and steepens the decline in $\gamma(w)$ with an increase in w at the smallest values of w (fig. 5.2B). This makes intuitive sense because as food density declines, any given food cluster size and resource concentration is encountered with lower probability. This means that species must be more separated in sampling scale to gain enough exclusive resources to meet requirements. Larger species thus do not need to be separated in scale as much as smaller species because of their larger total sampling volume V, and greater access to more total food per unit time. Unlike the case for resource concentration, decreasing the fractal dimension of food F does not cause $\gamma(w)$ to increase with w (fig. 5.2B).

A final observation on limiting similarity is that under conditions where many species can coexist, such as when food is dense (high F), resource concentration is high (high Q) and/or species interact at a large spatial extent x, scale ratios will be close to 1 and change relatively little over one or more orders of magnitude in sampling scale. To the extent that body size predicts sampling scale, diverse communities may have many pairs of species with very small, or even no difference in size. Such patterns, which may seem to indicate no size structuring of the community, could arise even in communities of trophically similar species (guilds) that are limited by and compete for their shared resource. Also, a subsample of species from such a community might lead one to claim, justifiably, that scale or size ratios are invariant, as predicted by Hutchinson (1959).

With this analysis we have now connected classical optimal foraging theory to classical resource competition theory to generate a new "spatial scaling" model of community structure that embraces heterogeneity and generates a result sought since the 1960s (MacArthur and Pianka 1966; MacArthur and Levins 1967; MacArthur 1969). The limiting similarity rule derived here provides a tool for translating information about food and resource densities and distributions that predicts scaling and abundance patterns in communities of trophically similar species. We explore these in the next section.

Shared Resources and Sampling Scale

Now that we have determined limiting similarity and thus γ, the ratio of sampling scales from the conditions for competitive coexistence, we can explore further the question of species abundance. To predict species' abundance, we can use the equation for equilibrium population size in the landscape (eqn. (5.4)). First we need to understand how available food cluster sizes and resource concentrations change with species' sampling scale, how these combine to determine shared resources, and then explore possible patterns in species' abundance.

Our first step is to recall from chapter 4 that near the extremes in sampling scales that can feasibly exist in the community (w_{min} or w_{max}),

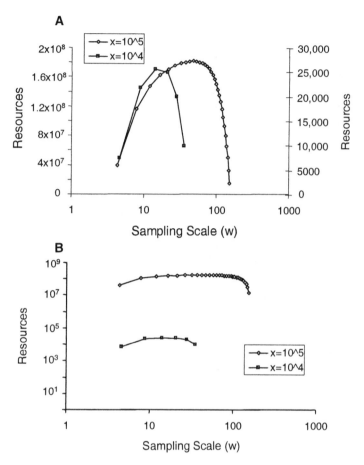

FIGURE 5.3. (A) Predicted change in abundance of resources (portion of shared plus exclusive) for species with different scales in two different-sized landscapes (different extents x), plotted on a linear scale. (B) Same relationships as in (A) but plotted on a logarithmic scale.

species' available resources decline dramatically (fig. 5.3A). By inspection of the equation for shared resources (eqn. (5.14)), we find that the smallest species are effectively limited by the availability of clusters with sufficiently high resource concentration, $R_{max}^{*\,(1+b)} - \sigma^{(1+b)}$ $(\gamma/w)^{(F-Q)(1+b)}$. The sampling scale dependent (right-hand) term subtracted from R_{max}^{*} is large for small values of w and the difference may even be negative. The smallest-scaled species are not limited by the availability of food clusters, given by $P_{max}^{*(1+a)} - \theta^{(1+a)}(\gamma w)^{F(1+a)}$ because the scale-dependent (right-hand) term subtracted from P_{max}^* is very small, implying that smaller-scaled species can exploit the full range of food cluster sizes available.

The converse is true for the largest-scaled species, whose large scale begins to reduce the range of available food clusters but allows them to use the full range of available resource concentrations. Effectively, smaller species are *concentration-limited* in resource availability, while larger species are *cluster size-limited*. The switch in major form of limitation means that resource availability is highest for intermediate-scaled species. This is most evident when examining resources plotted on a linear scale against log(w) (fig. 5.3A). However, plotting resources on a logarithmic scale (fig. 5.3B) shows that resources vary over only approximately an order of magnitude for feasible ranges of available food cluster sizes and resource concentrations. When compared to expected resource requirements, which may vary over several orders of magnitude, resource availability is thus approximately independent of scale.

This is an important result, because it provides a mechanistic justification for the "energy equivalence rule" (Damuth 1981, 1998, 2007), which states that different-sized species should access, on average, the same density of resources. Figure 5.3A clearly shows that, for all but the largest and smallest species coexisting in a community, resource availability is roughly constant relative to requirements, which I have assumed are proportional to $w^{-9/4}$ (or mass to the $-3/4$). Figure 5.3 also reveals that the scale of observation x strongly determines the pattern with which resource density changes with scale. At small scales of observation, the relationship between resource abundance and scale is more strongly hump-shaped, even on a log scale. At larger scales of observation, the relationship resembles more of a horizontal line in log-log space.

The alternative limitation of smaller-scaled consumers by resource concentration and larger-scaled consumers by food cluster size implies that, as predicted in chapter 4, there is a minimum- and maximum-scaled species that can exist in a given environment. The minimum scale is found by testing for the condition $I_{ijk} > 0$ under concentration limitation (eqn. (5.14)) and under the assumption that if the species has the smallest sampling scale in the community. Then γ must be 1 for that species (there are no smaller species, so $w_i = w_j$) and

$$w_{min} = [\theta^{(2/F - 1/Q)}\sigma^{-1/Q}]. \tag{5.18}$$

Likewise, the maximum scale occurs when $I_{ijk} > 0$ under cluster size limitation. For the largest-scaled species, γ should again be 1 because there are no larger species, so

$$w_{max} = c_P^{1/(F(1+F/D))} x^{1/(1+F/D)}/\theta^{1/F}. \tag{5.19}$$

As expected these two extremes in sampling scale depend on F and Q, with w_{min} being higher when resource concentrations are higher and more uniformly dispersed within food (Q closer to F), and in general when food is less abundant (F is lower). In contrast, w_{max} is larger when there is more food and it is more dispersed (F is larger) and, since c_p is a fraction, when resource concentrations are lower on average or are more clustered within food (smaller $|Q–F|$ in the parameter a).

An interesting result is that w_{max} increases with the scale of observation (or size of the landscape) while w_{min} does not. As I discussed in chapter 4, the maximum resource concentration, which determines concentration-limited w_{min}, is a function of R^*_{max}, the smallest possible food cluster size that still allows one cell to be occupied by resource. In contrast, w_{max} is determined by P^*_{max}, a function of the maximum food cluster size. Larger landscapes lead to larger possible aggregations of food (see fig. 4.6), by random clustering of occupied cells, regardless of their resource concentration, just because there is more possible space for such an aggregation to occur. This connection between extent x, P^*_{max}, and w_{max} for the community becomes very important when I explore predicted patterns of species diversity in chapter 6.

A testable qualitative prediction of the spatial scaling model of community structure is that w_{max} should increase with area A of a landscape. Following equation (5.19), and assuming a 2-dimensional landscape ($D = 2$) and therefore $x = A^{1/2}$, then $w_{max} \propto A^{1/[2(1+F/D)]}$. Given maximum and minimum possible values for F, the exponent of A can vary between 1/4 when $F = D$ and 1/2 when $F = 0$ (fig. 5.4). Interestingly, this suggests that maximum scale increases *more slowly* with area in more productive (higher F) landscapes. This occurs because productive landscapes tend to have larger clusters of food even at small

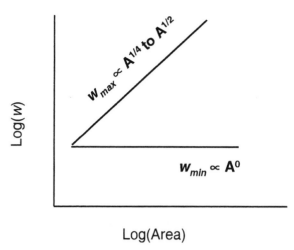

Log(Area)

FIGURE 5.4. Predicted scaling relationships for maximum scale (w_{max}) and minimum scale (w_{min}) vs. area of landscape, plotted on logarithmic axes. Note that a range of slopes (exponents of scaling relationships) are predicted for w_{max}, depending on F, but that the maximum slope should be 1/2.

extents, so increasing extent doesn't increase maximum food cluster size and thus w_{max} as much.

Shared Resources and Species' Abundance

The signature of "resource equivalence" among species with different sampling scales dominates the abundance structure of communities predicted by the spatial scaling model of species coexistence. Following equation (5.9), abundance of species j is a function of the shared resource I_{ijk} divided by its relative allocation among the next larger and next smaller species, plus the exclusive resource E_j, all divided by estimated resource requirements B_j. Calculations of N_j (fig. 5.5) reveal the following consistent pattern. The smallest species has the highest abundance, because its resource requirements are low compared to the amount of shared resources it can obtain, given the amount of exclusive resources it needs to guarantee persistence in the community. Abundance declines as sampling scale increases, slowly at first, because increases in scale proffer the use of a wider range of resource

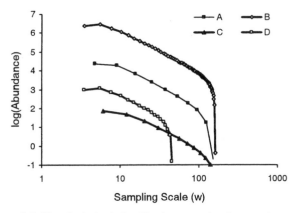

FIGURE 5.5. Hypothetical relationships between abundance and sampling scale predicted by the spatial scaling model of community structure. Curves reflect pattern, symbols represent predicted individual species' abundances. (A) $x = 10^5$, $F = 1.8$, $Q = 1.4$. (B) The consequences of higher $Q = 1.6$. (C) The consequences of lower food density $F = 1.4$, $Q = 1$ (density 1×10^{-3}, resource concentration 0.01). (D) The consequences of smaller extent ($x = 10^4$, F and Q as in (B)). Note this results in higher mean food density although mean resource concentration ($Q - F$) remains constant. The slope of the decline in abundance with sampling scale at intermediate scales is very nearly -2.25 when $D = 2$.

concentrations. And then at even larger scales, abundance declines rapidly, approximately with a slope ranging between -2 and -2.25 on a log scale [log N_j vs. log(w)]. At the largest sampling scales, abundance decreases very rapidly, as the range of suitable food cluster sizes begins to decline dramatically.

Abundance-sampling scale patterns are strongly affected by mean food and resource densities and distributions (F and Q respectively), and the extent of observation (x) (see curves A–D in fig. 5.5). Increasing resource concentration or its dispersion within food (increasing Q) increases overall abundances of species, but not the maximum or minimum sampling scale of a species (compare curves A and B). Decreasing food density or increasing its clustering (decreasing F) results in lower abundance but shows little change in maximum or minimum sampling scales of species (compare curves A and C). Finally, decreasing the extent of observation by one factor of 10 leads to a smaller maximum food cluster size in the environment, a lower maximum sampling scale for the community and, because of the reduction in the range of food cluster sizes available, a greater than 100-fold proportional reduction in abundance (compare curves B and D).

Figures 5.2–5.5 outline a rich set of qualitative predictions that can be tested on real communities. There are predicted relationships between scale ratios vs. scale, maximum and minimum scales vs. area of extent, and abundance vs. scale. In the next section, I will test some of these predictions with both data from the literature and a set of well-studied communities of trophically similar species, as is assumed by the spatial scaling model of community structure.

ABUNDANCE AND SIZE PATTERNS IN REAL COMMUNITIES

Ecologists have studied patterns of size and abundance for nearly 50 years, dating back to Hutchinson (1959) and Preston (1962). Most of these studies are unsuitable for testing the predictions of the spatial scaling model of community structure because they: (1) lumped all species of a similar taxonomic group, such as all birds, insects, or mammals regardless of their trophic status; (2) lumped species of very different resource consumption, such as seeds vs. plant leaves, into single trophic categories, like herbivores; (3) did not measure abundance of species; or (4) sampled communities that were almost certainly missing species because of excessive disturbance. Consequently, there are surprisingly few data available from communities for testing the model.

Here I combine data of my own that have not yet been published with a few published studies of size and abundance for clearly defined trophic groups. The published data include Marquet and Taper's (1998) data on maximum and minimum sizes of mammals, and John Terborgh and colleagues' (1990) extensive sampling of birds in a Peruvian rainforest. Other data come from my own studies.

The first unpublished data set is a compilation of ungulate censuses in 2001 from Serengeti National Park (Anderson et al. 2007), plus studies of local densities of smaller mammals, such as hyraxes (*Heterohyrax brucei*, *Procavia capensis*), and klipspringers (*Oreotragus oreotragus*), that were not included in the censuses (Hoeck 1989; Dunbar 1990; Mahlaba and Perrin 2003). These three species are obligate associates with a specialized habitat, granite outcrops, in the

Serengeti and I multiplied their reported local densities by the estimated area of habitat. Censuses were counts of the number of animals of each species observed from a Land Rover in 15 belt transects 200 m wide and 10 km long dispersed throughout the park. I multiplied the mean density of a species from all transects by the size of Serengeti National Park (12,700 km²). Sizes of species were taken from Kingdon (1984).

The second unpublished data set is dung beetles (Coleoptera: Scarabeidae) sampled from Hluhluwe-Umfolozi Game Preserve in KwaZulu Natal, South Africa. In April 2002, I sampled fresh (<4 hrs old) dung of various large mammalian herbivore species including elephant (*Loxodonta africana*) white rhino (*Ceratotherium simum*), buffalo (*Syncerus caffer*), wildebeest (*Connochaetes taurinus*), giraffe (*Giraffa camelopardalis*), and impala (*Aepyceros melampus*) found in shrub and grassland habitats in the reserve. If ball-rolling dung beetles (Hanski and Cambefort 1991) were present, I collected as many as possible by hand and with a butterfly net as soon as I approached the dung pile. I then measured the volume of the dung pile from its mean diameter and thickness. I sorted completely through each fresh dung pile and extracted all beetles >5 mm in size and then deposited a measured volume of remaining dung containing beetles <5 mm into a plastic bag. Extracted beetles and the dung subsample were immediately placed on ice and then frozen at −15°C later the same day. I later thawed the samples and sorted extracted beetles to species and counted them. I then extracted, sorted to species, and counted all beetles <5 mm in size from the dung subsample. All identifications were made on the basis of morphospecies distinctions that were later corroborated by official identification in a reference collection by Adrian Davis of the South African Museum of Zoology in Pretoria. Once identifications were known, I measured mean length in mm of up to 10 individuals (males and females mixed) of each species.

I also used another insect dataset, this time for a guild of foliage-feeding herbivores, the grasshoppers (Orthoptera) at Cedar Creek Natural History Area in Minnesota (see Ritchie 2000). I measured the density and species composition of grasshoppers in an old field prairie from a sequential catch-effort method (Ritchie 2000) in August of 2005. Grasshopper identifications were confirmed against a Cedar Creek reference collection developed by entomologist John Haarstad. I measured mean live mass for up to 10 individuals for each species from fresh sweep samples in the same field. Many species occurred as nymphs, and I used the size of the species in the sample rather than the adult size in the analysis.

I also developed another dataset on plant size and abundance in the same old field prairie in 2005. Ten 0.5 x 1 m subsample frames were placed over the vegetation in each of four 100 m^2 plots (40 plots total). Field crews estimated visual percent cover of each species of a wide variation in plant types from mosses (not identified to species) to tall warm season (C_4) grasses. As a measure of plant size, I measured the canopy width (mean distance from outermost plant part to the opposite plant part for four different distances) for each of 10 individuals of each of the species found in the percent cover plots. I then divided the summed estimated cover for a species over all sampling frames by the canopy cross-sectional area estimated from mean canopy width for a species.

The data obtained from the literature for Peruvian rainforest birds (Terborgh et al. 1990) required some filtering to assure that the species analyzed were from the same feeding guild. Although Terborgh et al. (1990) assigned every species a mass and a feeding guild, I confirmed the mass and feeding guild of each species from various online biodiversity databases, such as http://www.worldwildlife.org/wildfinder/. Plus, a similar study of birds in Panama (Robinson et al. 2000) provided additional data on mass and diet preferences that I used to adjust the measured mass (the mean of values reported in Terborgh et al. (1990), Robinson et al. (2000), and websites) and guild classification for some species. I also eliminated bird species that had no reported density or were sampled with a different census method. I had difficulty in accepting all guilds assigned by Terborgh et al. (1990), as some seemed over-inclusive, such as "gleaning insectivores" that included many ant specialists or "arboreal omnivores" that may have included species groups that ate very different things. Other classifications seemed overly specific, such as insectivores specialized on gleaning dead leaves. I also did not use guilds, such as nectarivores (hummingbirds), that had less than 8 species for which densities were recorded, since such guilds provide little statistical power for testing the model. In the end, I used only two guilds for which I had confidence that all species actually ate similar things: woodpeckers and bark gleaners (Picidae, Dendrocolaptidae and Furnariidae) and raptors

(Strigidae, Accipitridae, Falconidae). Reported abundances were the observed number of breeding pairs for each species in these two groups multiplied by 2.

A final guild I used to test quantitative predictions of the spatial scaling model was obtained from Dolan et al. (2007) for a group of protist ciliate herbivores called tintinnids, which are diverse basket-shaped micro-zooplankton (30–350 μm in length) that feed on algae and bacteria in pelagic ocean environments (Dolan 2000). I used a single filtered sample from a cross-latitudinal survey in the Pacific Ocean conducted in 2005 that featured 18 species of tintinnids and chlorophyll (algal) concentrations of 0.41 ppb in a 60-liter sample.

Maximum and Minimum Size in Communities

A major qualitative prediction of the spatial scaling model is that there should be maximum and minimum sizes found in a community of trophically similar species. It should also predict how they change with the volume or area of space in which the community lives. The only data I am aware of that test for what controls the maximum and minimum sizes in a community is that of Marquet and Taper's (1998) macroecological study of the smallest and largest mammals found on islands and continents of different size (fig. 5.6). To compare the quantitative predictions of the spatial scaling model with the observed pattern, note that equation (5.19) implies a scaling relationship between mass and area by assuming that $w \propto M^{1/3}$. If so, then equation (5.19) suggests $M_{max} \propto A^{3/[2(1+F/D)]}$ and as $F \to D$, $M_{max} \propto A^{3/4}$. The observed data for maximum size are fit well by a scaling relationship with increasing island or continent area, as predicted by the spatial scaling model. However, the observed exponent or slope of the log-log regression of 0.54 (±0.04 SE) is significantly less than the predicted minimum exponent of 3/4 (fig. 5.6).

This discrepancy in predicted vs. observed slopes might be explained in a number of ways. There could be additional constraints on maximum size not accounted for by the spatial scaling model, such as the possible failure of larger species to disperse to islands that are large enough to sustain them. The impact of humans on extant fauna cannot be underestimated, since the arrival of high population densities of

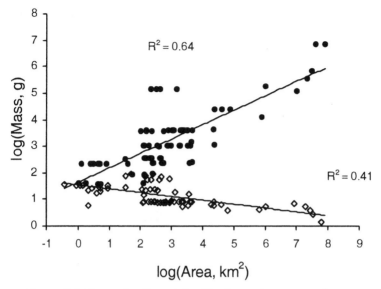

FIGURE 5.6. Observed scaling relationships for maximum mass of mammals (a surrogate for w_{max}), and minimum mass of mammals (a surrogate for w_{min}) reported for different sets of islands and continents of different land area, plotted on logarithmic axes. The equation for maximum mass, M is $\log(M) = 1.64 + 0.54 \log(A)$, and the equation for minimum mass is $\log(M) = 2.1 - 0.193 \log(A)$. Data from Marquet and Taper (1998).

humans on several continents and islands apparently coincided with the extirpation of the largest mammals, causing the observed current maximum size of mammals found on continents to be lower than true capacity. Or, the model could just be plain wrong. However, the spatial scaling model is the only formal model that explains why maximum size should scale with area, and its predicted exponent is at least near the observed exponent, even if it is significantly different.

The spatial scaling model predicts that minimum size should be independent of extent x and thus of area A (eqn. (5.18)). However, for mammals, minimum size decreases as a scaling relationship with increasing area with an exponent of -0.19 (± 0.02 SE), which is significantly less than zero. This discrepancy suggests that some aspect of the environment changes with increasing area in a way to favor smaller minimum size. One hypothesis is that the smallest islands lack sufficiently high quality habitats or food, or lack food with sufficiently high resource concentrations. Alternatively, the smallest species may have difficulty colonizing small islands where they otherwise would have the capacity to persist.

In both of these observed scaling relationships, data are from a mixture of trophic groups. For minimum size, the smallest species is likely to be a bat, shrew, or in some cases a rodent, and the occurrence of new taxa across islands or continents might be sufficient to explain the negative scaling exponent. For maximum size, smaller islands may lack larger herbivorous mammals because they have difficulty colonizing or are more likely to go extinct. In either case, and for other reasons, as well, shifting presence of different trophic groups across islands of different size may introduce noise that weakens the regressions and tends to reduce slopes, perhaps enough to cause disagreement between theory and prediction.

The discrepancy between theory and observation suggests there is much research needed on what controls the smallest sizes within communities. Given that there are few hypotheses and even less carefully analyzed data that address the subject, this is an area that is ripe for further research.

Size Structure of Communities

The second major pattern to explore is the size structure of a community of trophically similar species. This includes both how far apart in size species should be and the abundance of each species. Again, as in previous chapters, I assume that sampling scale $w \propto L$ and $w^3 \propto M$ where M is mass. The spatial scaling model predicts that the size ratio should be highest for the smallest size pair and then decline rapidly to an asymptote at the largest size pairs (fig. 5.2). The pattern is most visible when the ratio is plotted versus the logarithm of size. The model also predicts that the logarithm of abundance should decline nonlinearly with increasing logarithm of scale or size, with a long interval of sampling scales or body sizes over which the decline appears linear with a slope of approximately -2.25 vs. log(length) or $-3/4$ vs. log(mass) (fig. 5.5). These asymptotic slope values reflect the strong signal of the assumption that resource requirements scale as $L^{9/4}$.

I fit the fractal model to the data for each trophic guild by a 3-step process. The fractal model has six parameters if you fix the exponent for the scaling of resource requirements to be 9/4: extent x, food fractal dimension F, resource fractal dimension Q, the background dimension of the environment D, the number of sampling volumes v, and the pre-factor β_0 for resource requirements. Data are available to estimate β_0 for mammals, insects (at standard temperature), and birds (Peters 1983; Anderson and Jetz 2005). I knew the extent (in mm) of the landscapes in which each community was sampled, so this provided x.

The background dimension D was not a constant value of 2, because different guilds operated in environments with different degrees of 3-dimensional structure. The influence of such structure on each guild is obtained by assuming that the area was isotropic, that is equal in length and width and thus $= x^2$. The vertical component of height y can be expressed as a fraction δ of x, where $\delta = y/x$. By our previous equations converting proportion occupied into a dimension, a vertical dimension Y is

$$Y = 1 + \ln(\delta)/\ln(x) . \tag{5.20}$$

Adding Y to the width and length dimensions yields $D = 2 + Y$. So the additional height of trees for bark-feeding birds, the height of grass for grasshoppers, the depth of roots for plants, and of course the fully 3-dimensional pelagic environment of tintinnids resulted in adjustments of D described in table 5.1. Model parameters were obtained by a priori assignment and least-squares best fit of the spatial scaling model to the relationship of size ratios vs. size of the larger of each ranked pair for 7 different feeding guilds (see figs. 5.7–5.10).

This leaves three parameters that were fit through non-linear regression to the size ratio data. The second step in fitting the model is to make initial guesses for F and Q based on approximate proportion of the landscape occupied by food m and the proportion of food occupied by resource q/m (see chapter 4)

$$m = x^{F-D} q/m = x^{Q-F}. \tag{5.21}$$

Thus

$$F = D + \ln(m)/\ln(x) \quad Q = F + \ln(q/m)/\ln(x) \tag{5.22}$$

TABLE 5.1. Model Parameters for 7 Different Feeding Guilds

Community	S	x (mm)	F	Q	D	v	BI	$\gamma_{null} \pm SE^{\#}$
Mammal herbivores	19	1.3×10^8	1.85	1.37	2.0	1320	78.4	1.75 ± 0.24
Dung beetles	28	2000	1.88	1.66	2.5	1000	39.0	1.06 ± 0.01
Grasshoppers	16	9000	1.90	1.47	2.2	800	39.0	1.31 ± 0.14
Plants	24	16,000	1.67	1.21	2.3	3800	7.0	1.19 ± 0.02
Bark-feeding birds	21	4.5×10^6	1.38	1.23	2.2	700	78.4	1.17 ± 0.07
Raptors	15	4.5×10^6	1.32	1.22	2.0	800	78.4	1.14 ± 0.04
Protist herbivores	18	400	1.61	1.19	3.0	2000	35.6	1.21 ± 0.04

Mean ratio (±SE) is shown from a null model of species sizes obtained from a random uniform distribution of log(size), γ_{null}.

The final step alters the parameter v, which is left to be the only fully free parameter, to be fit in the customary way with non-linear regression, using occasional adjustments in F and Q.

To test whether the data could be equally fit with a random model, I developed a simple null model. For a given number of species S, I selected $S - 2$ log(size) values from a uniform random distribution of log(size) with observed minimum and maximum log(size), ranked them, and then calculated the size ratio of all sequentially ranked pairs. A uniform distribution is much more likely to fit a distribution of values purposefully dispersed along an interval than a lognormal or other modal distribution. I did this 100 times for each guild and calculated a mean and standard error. There was a small sensitivity to number of species, where less diverse guilds had higher mean size ratios. However, I found that there was no expected relationship between mean size ratio and log(size) for the null model. To more rigorously test the spatial scaling model against the null model, I calculated Akaike Information Criteria (AIC) for each model for each guild, which calculates a fit to the data penalized for the number of parameters.

Employing this process for each guild, I found that the spatial scaling model fit the size ratio data fairly well for six of the seven datasets

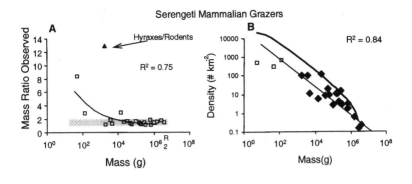

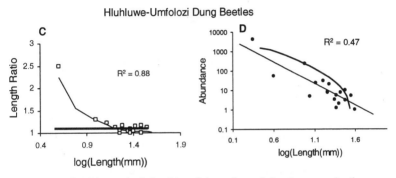

FIGURE 5.7. Observed relationships of size ratios and abundance vs. size for two guilds in tropical savannas. (A, B) Foliage-eating mammals in the Serengeti National Park, Tanzania from the Serengeti biodiversity project (Anderson et al. 2007). (C, D) Dung beetles found in white rhino dung in Hluhluwe-Umfolozi Game Preserve, South Africa from my own unpublished data. In (A) and (C), the thin curved line is the fitted (least squares and maximum likelihood) spatial scaling model; the gray band or line is the mean size ratio (±SE) from a uniform random null model of sizes for $S-2$ species between the observed maximum and minimum sizes in each community. In (B) and (D), the thick curve is the predicted abundance from the spatial scaling model fitted to the pattern for size ratios; the thin line is the energy equivalence model: $\log (y) = b_0 - 3/4\log(x)$ fitted to the abundance data. See table 5.2 for model fit comparisons.

of trophically similar species or guilds. For these six guilds (figs. 5.7–5.9), the qualitative predictions of the spatial scaling model are supported for the *size ratios* of species. For Serengeti mammalian herbivores, dung beetles, plants, and rainforest raptors, fits for size ratios vs. size were strong, with $R^2 \geq 0.75$. Fits for grasshoppers and bark-feeding

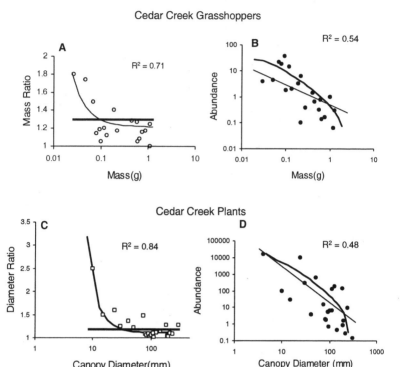

FIGURE 5.8. Observed relationships of size ratios and abundance vs. size for two guilds in temperate grassland of an abandoned agricultural field at Cedar Creek Natural History Area, Minnesota. (A, B) Grasshoppers (Orthoptera: Acrididae, Tettigoniidae) the dominant insect folivores, and (C, D) herbaceous plants, data from Ritchie (2000). Curves shown are the same as those described in fig. 5.7.

rainforest birds were somewhat weaker $(0.5 < R^2 < 0.75)$, but still highly significant. In these six cases, AIC scores were very much lower for the spatial scaling model than the random model. Consequently, for a variety of different guilds, ranging in size from 10 mg to 3000 kg, the spatial scaling model appears to predict the qualitative and quantitive size structure.

For the ciliate protists (fig. 5.10), the model did not predict the size structure well at all; in fact, a hypothesis of random variation in size fit better (although not particularly well). The random model had a higher R^2 and lower AIC, suggesting that size structure of the protist

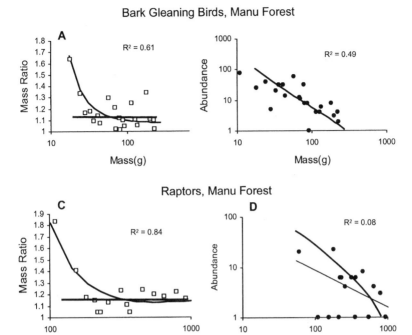

FIGURE 5.9. Observed relationships of size ratios and abundance vs. size for two guilds of birds in a tropical rainforest, the Manu Forest in Peru. (A, B) Bark-feeders, including woodpeckers (Picidae) and woodcreepers (Dendrocolaptidae), and (C, D) raptors, both nocturnal and diurnal. Data from Terborgh et al. (1990), with body masses averaged from other sources. Curves shown are the same as those described in fig. 5.7.

community was much more likely to be the result of a random collection of species. Interestingly, $D = 3$, and $F = 1.6$ was estimated from measured algal concentrations, which means that food was by far at its lowest mean concentration for these protests. Food at concentrations this low may be sufficiently scarce that relatively non-vagile consumers like these ciliates may have had difficulty encountering food items in v search volumes, regardless of their cluster size or resource concentrations. Therefore, my assumption to this point that p_F, the probability of encountering food in a search volume of size w, is equal to 1 is probably incorrect. Indeed, calculations of p_F for the range of likely w (length) values for tintinnids yields $p_F < 0.02$. This may mean that

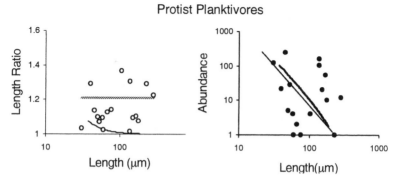

FIGURE 5.10. Observed relationships of size ratios and abundance vs. size for a guild of planktivorous protist ciliates, called tintinnids, of the pelagic South Pacific Ocean (Dolan et al. 2007). Theoretical and fitted curves shown are the same as those described in fig. 5.7.

for ciliate protests and other small, non-vagile species consuming scarce food (Gotelli and Rohde 2002), community dynamics may be strongly dispersal-limited in their composition and diversity and de-scribed better with neutral theory (Hubbell 2001) or statistical state variable models (Harte et al. 2005, 2008).

Observed species abundances in 5 of the 7 guilds qualitatively were decreasing scaling relationships with increasing size, as predicted by the spatial scaling model. However, there was considerable scatter around the relationships predicted by the model that best fit the size ratio structure of each guild. Table 5.2 compares the spatial scaling (SS)(K=3) and random (RAND)(K=1) models for size ratios vs. log(size) and for the SS model and EER (K=1) for log(abundance) vs. log(size) for 6 guilds. R^2 was obtained from a least-squares regression and Akaike maximum likelihood criteria (AIC). A smaller (more negative) AIC value indicates a better model fit to data after correcting for K. In most cases, the model seemed to predict the upper limit to abundance (fig. 5.7, 5.8), and most of the scatter resulted from many smaller species with abundances much lower than that predicted by the model. In addition, the data were even better fitted by a line (in log-log space) with a fixed slope of –3/4 for mass data and –9/4 for length data but fitted intercept, as predicted by Damuth's energy equivalence rule (EER) (table 5.2). Largely be-cause EER had only one fitted parameter (K) while the spatial scaling model had 3, and because EER was fit to abundance data rather than

TABLE 5.2. SS, RAND, and EER Model Comparisons

| Community | S | Size Ratios | | | | log(Abundance) | | | |
| | | SS | | RAND | | SS | | EER | |
		R^2	AIC	R^2	AIC	R^2	AIC	R^2	AIC
Mammal herbivores	19	0.75#	−7.25	0.008	19.1	0.81	−34.1	0.78	−33.5
Dung beetles	28	0.89	−58.6	0.021	−25.1	0.50	−16.1	0.71	−27.9
Grasshoppers	16	0.71	−71.5	0.003	−53.0	0.58	−23.2	0.56	−24.4
Plants	24	0.84	−88.8	0.013	−44.6	0.42	5.6	0.47	1.6
Bark-feeding birds	21	0.61	−95.7	0.054	−80.8	0.51	−41.6	0.50	−43.4
Raptors	15	0.82	−70.1	0.021	−46.3	0.05	−22.0	0.08	−23.5
Zooplankton	23	0.00	−42.7	0.108	−60.2	0.08	−17.6	0.10	−20.4

Excludes size ratio of hyraxes to rodents, since mammals of the expected intermediate size are uncommon and not folivores in Africa, even though they are common on other continents, especially North America.

size structure data, AIC values for the EER hypothesis were consistently lower than that for the spatial scaling model.

The two poorest fits for the spatial scaling model and EER were rainforest raptors and the pelagic ocean ciliates. The rainforest raptors had several species represented by a single individual, indicating only partial or occasional use of the studied area by these species (fig. 5.9). If these species were excluded, likely both the spatial scaling model and EER hypothesis would have fit abundance data much better. This again suggests that the fractal model predicts an upper limit to abundance. For the protists, there was very weak correlation between abundance and size, which was consistent with the fact that the community exhibited little size structure. If dispersal limitation and community "drift" are the dominant forces structuring species abundance, then a close fit with the resource-based spatial scaling and EER models would not be expected.

The close correspondence of the spatial scaling and EER models in explaining the abundance data suggests several overriding conclusions. With the exception of the protist ciliates, the abundance structure of the communities examined matches qualitatively with the predictions of the models. The most abundant species tend to be the smallest and the least abundant species tend to be the largest. While there is considerable scatter in the abundance of individual species of particular sizes, the spatial scaling model seems to describe an upper bound to abundance. It is easy to imagine that dispersal limitation, differential tolerance of local environmental conditions, predators, diseases, or other factors might drive abundance of a particular species below the limit predicted by the spatial scaling model. Clearly, however, these guilds do not seem to have the modal abundance vs. size curves predicted by the reproductive power hypothesis (Brown et al. 1993; Maurer 1998), or the dispersal vs. metabolism trade-off hypothesis (Etienne and Olff 2004), or even the central limit theorem for multiple limiting factors (May 1975; Siemann et al. 1996). The predicted patterns of the model and data come very close to the patterns predicted by a fractal resource abundance hypothesis of Morse et al. (1985) (slope of −9/4 in log abundance vs. log length) and the observed patterns of decline in abundance of rainforest insect communities. However, the patterns of the spatial scaling and EER predictions are driven by the mechanism that resource availability is *similar across sizes* rather than *declining as $L^{-9/4}$* and that it is the allometric scaling of *resource requirements* (West et al. 1997; Brown et al. 2004) that causes abundance to scale as $M^{-3/4}$ or $L^{-9/4}$, not the scaling of available habitat or resources. A recent study of aquatic invertebrates inhabiting the complex structural forms of macrophytes (McAbendroth et al. 2005) provides additional support for this conjecture, as size-abundance relationships from macrophyte species that differed greatly in their structural complexity all declined as the 3/4 power of mass.

The close correspondence between quantitative predictions and fit of the spatial scaling and EER models suggests that heterogeneity and the trade-offs organisms experience to accommodate it are a mechanism for why "energy equivalence" might be a viable hypothesis to explain patterns in abundance of species of different size. The spatial scaling model predicts that resources available from exclusive and shared resources will vary by less than an order of magnitude across sizes except for the very smallest and largest species, with a modal peak at intermediate size. If mass varies across 3–4 orders of magnitude within a guild, then the signal of resource requirements and their allometric scaling will likely be much stronger than variation in resource availability with size. Thus, the trade-off between smaller species needing food clusters with higher resource concentration vs. larger species needing larger food clusters leads to this approximate convergence in resource availability, and therefore a mechanism to explain "energy equivalence" for species of very different body size.

A third outcome is that declining size ratios with increasing size is a robust qualitative prediction that is supported by the limited set of guilds I explored here. However, such a pattern of size structure is not guaranteed statistically, as the pattern for protist ciliates clearly did not conform, and their abundance did not scale significantly negatively with body size. The contrast between model fit for some guilds and complete lack of fit for others suggests, following Rohde (2001) and Gotelli and Rohde (2002), that there may exist many different types of guilds. Those with highly vagile species that can encounter food and resource clusters with fairly high probability, such as vertebrates, mobile insects, and/or plants may approach potential density saturation and have communities structured by interspecific competition. Those with species that have poor vagility and/or extremely low density resources, such as Dolan's (2000) protist ciliates or Klaus Rohde's (2001) fish parasites, may be structured much more by dispersal limitation and have relatively random size structure.

This leads to the fourth major outcome, that the spatial scaling model presents a clear alternative *niche-assembly model* to neutral or dispersal-limited models of community assembly (Tilman 1994; Hubbell 2001) and to classical consumer-resource models (MacArthur 1969; Tilman 1982; Chase and Leibold 2003). Its patterns are quite distinct, with qualitative predictions of non-linearly declining size ratios with increasing size, and declining abundance with increasing size. Most other community assembly models (MacArthur 1969; Tilman 1982; Chase and Leibold 2003; Hubbell 2001; Volkov et al. 2003; Harte et al. 2008) make no predictions about size. Those that predict size patterns (May 1975; Morse et al. 1985; Brown et al. 1993; Siemann 1996) have no explicit assembly rules or mechanisms and make no predictions about community membership. The spatial scaling model of community structure is also the only model available that incorporates spatial heterogeneity explicitly (described with fractal geometry). With its unique processes and predicted community patterns, it seems to be a strong hypothesis for community structure in a variety of contexts.

SUMMARY

1. Numerous hypotheses exist to predict or explain the pattern in abundance of species vs. body size, but few of these are linked to explicit mechanisms of community assembly or coexistence. Ecologists still do not have many viable hypotheses to predict the maximum and minimum sizes, size similarity, and the abundance of different-sized species in communities.

2. The fractal niche model for food cluster size and resource concentration presented in chapter 4 predicts the size structure of communities of guilds, or species that are limited by similar resources. The niche model is combined with previous dynamical models of competition for shared and exclusive resources (fig. 5.1) to predict the *magnitude of shared and exclusive resources* for species of different sampling scale and *species packing*, or how close in scale species can be and still persist in the community.

3. The resulting *spatial scaling model* of community structure for guilds predicts that, under a wide set of conditions, species with smaller sampling scales (size) should be farther apart in scale (size) than species with larger sampling scales (size) (fig. 5.2). The combination of shared and exclusive resources available for species with different sampling scale is unimodal with a peak at intermediate scales (fig. 5.3A), but these availabilities vary by less than an order of magnitude over 1.5 orders of magnitude in length or 3–4 orders of magnitude in scale (fig. 5.3B). These results combine to predict that smaller-scaled species should be more abundant than larger-scaled species (fig. 5.5) and that abundance should be inversely proportional to resource requirements.

4. The spatial scaling model also predicts a maximum and minimum sampling scale (size) for a guild. The maximum scale (size) is predicted to increase as a scaling relationship with increasing extent (area) of the "landscape" in which species are coexisting, while minimum scale (size) is predicted to be independent of area (fig. 5.4). A published study of minimum and maximum body masses of mammals on islands and continents (Marquet and Taper 1998; fig. 5.6) supports the qualitative predictions of the model for maximum

size, although the slope of the scaling relationship with area is less than predicted. In addition, the minimum mammal body mass declines significantly, but very gradually, with increasing area. The whole topic of minimum and maximum sizes needs much more theoretical and empirical work.

5. The theoretical predictions in item 3 above were tested with seven different guilds of 15–25 species ranging in size from terrestrial mammalian herbivores to protist ciliates in pelagic ocean environments. For 6 of the 7 guilds, patterns in size ratios of adjacent-sized species were fit very well by the spatial scaling model (figs. 5.7–5.9), and much better than a null model of random body size. For the protist ciliates, body size structure was more consistent with a random model than the spatial scaling model (fig. 5.10).

6. The quantitative predictions of the spatial scaling model, fitted to the size structure but used to predict abundance vs. size, were compared with an alternative model, that of "energetic equivalence" (EER) (Damuth 1991). In 5 of the 7 guilds studied, both models yielded similarly strong quantitative predictions and fits to data, with the smallest species being more abundant than the largest species (figs. 5.7–5.9). The similarity in predictions between EER and the spatial scaling model was expected because the spatial scaling model predicted resource availability to vary much less with size than resource requirements and thus, on logarithmic scales, to be "equivalent" among different-sized species. For the other two guilds, many smaller species were as rare as the largest species, weakening correspondence with the models (figs. 5.9, 5.10).

7. Overall, the spatial scaling model yields a set of highly testable hypotheses about the size structure of guilds that can be compared with various neutral or null models or with alternative hypotheses, such as EER or the reproductive power hypothesis. It seems to match the pattern of size-similarity in real guilds and predict an upper limit to abundance of species and the slope of the relationship between abundance and size. This success, for some guilds at least, suggests that the manner in which species partition heterogeneous distributions of the *same resource* can predict the size structure of communities and explain the coexistence of many more species than classic consumer-resource models.

CHAPTER SIX

Heterogeneity and Patterns of Species Diversity

Predicting species diversity and its major patterns from underlying mechanisms of organism births, deaths, dispersal, and resource consumption, to name a few, has perhaps been the "Holy Grail" of community ecology since the 1950s. Beginning in the 1960s, classical Lotka-Volterra models of competition and predation (Lotka 1925; Volterra 1926) published 30 years earlier began to be employed to understand how species coexist. By the beginning of the 1970s, ecologists were excited about constructing "community matrices" of interactions among sets of competitor or predator-prey species to predict species diversity and its patterns across different environments (MacArthur 1969, 1972; Vandermeer 1972; Strobeck 1972). The bloom on this "rose" died with the understanding that matrices for communities of >4–5 species often predicted unstable dynamics and community composition (Strobeck 1972; May 1974). The later evolution of Lotka-Volterra models as "consumer-resource" models properly recognized that the competition coefficients of these models reflected not differences in dietary overlap, but rather, differences in the rate of consumption of one or more potentially limiting resources (Tilman 1976, 1982; Hsu et al. 1977). Such consumer-resource models remain to this day the most popular for exploring the consequences of species interactions for community structure and diversity (Chase and Leibold 2003; Amarasekare et al. 2004).

The crux of the problem with consumer-resource models is that they predict that species coexistence is unlikely. Different versions, such as Tilman's (1982) models of competition or Chase and Leibold's (2003) models of predator-consumer-resource interactions, predict that any two species can coexist only under a relatively narrow range of re-source (or predator) supply rates. Diversities typically observed in na-ture can only be obtained by assuming that there is "heterogeneity" in supply rates, which is never explicitly mathematically modeled, but usually just depicted graphically as a "set" of possible supply rates across space in which community dynamics will be measured. Even with such suggested variation, it is still often difficult to get more than 7 or 8 species to coexist.

Recent models suggest that consumer-resource dynamics can gen-erate much higher levels of diversity when there is temporal variation in resource supply (Sommer 2002) or temporal variation driven by non-linear (chaotic) dynamics of the consumer species (Huisman and Weissing 1999, 2001). Other models, such as competition-colonization trade-off models (Tilman 1994; Pacala and Tilman 1994; Klausmeier and Tilman 2002) allow virtually infinite numbers of species to coex-ist, but it is dispersal limitation that effectively allows coexistence, be-cause species occupying the same location in space are assumed to fail to coexist. Further variants in which dispersal limitation plays a key role in coexistence are metacommunity models (Amarasekare and Nis-bet 2001; Amarasekare et al. 2004; Leibold et al. 2004), in which spe-cies interact in discrete local environments connected by dispersal at various probabilities among multiple local environments, which col-lectively constitute the "metacommunity." In these models, coexis-tence occurs from the way in which species over generations exploit heterogeneity in local conditions, over many localities (Klausmeier and Tilman 2002; Amarasekare et al. 2004).

Such models still ignore the fact that multiple species that use the same resource can still coexist in the same local environment, even without dispersal limitation (for example, Schoener 1983; Connell 1983; Belovsky 1997; Adams 2007). Such coexistence may be strongly driven by how different species exploit "local" heterogeneity *within a generation*, which presumably leads to differences among species in resource partitioning, habitat specialization, relative vulnerability to predators, etc. Despite the intuition that such heterogeneity should contribute to coexistence, the only model available to predict the coex-istence of species on local heterogeneity is that of shared and exclusive resources (Schoener 1976, 1978) and its new variant, the spatial scal-ing model of community structure that I introduced in chapter 5.

Another limitation of previous niche-based models of community structure is that they have not been used to predict many of the community patterns that ecologists routinely study. For example, classical consumer-resource models predict qualitative changes, that is, either increases or decreases in species richness with changes in resource supply; but only if the spatial variation in resource supply is known can the model predict the magnitude of change in species richness. This is a major limitation of current niche-assembly models because there is no predicted community structure to be compared with data or alternative models. For example, as far as I am aware, there is no current explicitly predicted species-abundance curve or diversity-size distribution for a model of niche assembly. Consequently, the role of niche assembly in structuring communities tends to be downplayed (see discussion in Hubbell 2001, chapter 1).

In this chapter, I apply the spatial scaling model of community structure (developed in chapter 5) to predict species richness and its major patterns and relationships between diversity and abundance. These patterns are often unique compared to predictions of neutral or dispersal-limited community models, and so the model provides a rich set of hypotheses for testing with field data. I explore some of these quantitative predictions with data from the 7 guilds I used in chapter 5, with many interesting results.

HETEROGENEITY, SCALE, AND SPECIES DIVERSITY

It was likely obvious in the preceding chapter that the spatial scaling model, in addition to predicting size structure, also predicts species richness. In fact, the model predicts a wealth of commonly described diversity patterns for ecological guilds, including diversity vs. size, area, spatial heterogeneity and productivity. First, I show the equation for calculating species diversity implied by equations from chapter 5 for maximum sampling scale w_{max} (eqn. (5.19)), minimum sampling scale w_{min} (eqn. (5.18)) and species packing rule $\gamma(w)$ (eqn. (5.16)). The basic idea is that, beginning at scale w_{min}, one can multiply $\gamma(w)$ consecutively $S - 1$ times to arrive at the maximum scale w_{max}. In formal mathematical terms,

$$\prod_{j=1}^{S-1} \gamma(w_j) = w_{max}/w_{min} \qquad (6.1)$$

where the index j refers to the rank of the species in sampling scale or size. This equation cannot be solved explicitly for S because $\gamma(w)$ is not a constant, but it is easily solved numerically for a given set of x, F, Q, D, v, and B_1.

I derive a general, fully parameterized expression for species richness by using a couple of additional mathematical approximations in box 6.1.

$$S = 1 + [F(1 + a)/(\theta^{\delta/2F}\sqrt{k})]\ln(c_P^{\phi/F}\theta^{(1/Q-1/F)}\sigma^{1/Q}A^{\phi/2})c_P^{\delta\phi/2F}A^{\delta\phi/4} \qquad (6.2)$$

However, this general equation isn't terribly instructive, as it hides the contributions of three parameters, Q, F, and D in parameters a, b, θ, σ, δ and ϕ; and hides the inverse relationship with resource requirements B_1, and positive relationship with the number of volumes sampled during the resource renewal interval v in the parameter k. A better approach in my view for understanding this multifactorial equation for species richness is to examine how the expression for S depends on variables that reflect major patterns of diversity with different independent variables, such as area, productivity, habitat fragmentation, etc. In the subsections below, I re-write equation (6.4) in various ways to illustrate the qualitative form of mathematical relationships for the major species diversity patterns. I keep parameters that are not related to the key independent variables as constants k_i to reduce the visual distraction of the large number of expressions in the model.

Species-area Relationships

The spatial scaling model predicts the following qualitative relationship for species richness vs. area

$$S = 1 + k_1 A^{\delta\phi/4}\ln(k_2 A^{\phi/2}). \qquad (6.4)$$

This is a very interesting prediction because it *does not predict a simple scaling law for species richness vs. area*, that is, it does not predict the familiar and virtually canonized $S = aA^z$, where a and z are constants, expected for species-area relationships (Rosenzweig 1995). This means that a niche-assembly process of partitioning resources by food cluster size and resource concentration *should not*, per se, yield scaling law species-area relationships. However, as A increases, the

BOX 6.1

A GENERAL ANALYTICAL SOLUTION FOR SPECIES DIVERSITY

Although the spatial scaling model does not yield a direct analytical solution for species diversity (eqn. (6.1)), a close approximation can be
solved. Beginning with equation (6.1), I assume that, as is true under
many conditions (chapter 5), $\gamma(w)$ declines rapidly to near an asymptote
$\gamma(w_{max})$, over the majority of species' sizes. The product of $\gamma(w_j)$'s from
$j = 1$ to $S - 1$ is thus approximately the product of $S - 1$ ratios for each
$\gamma(w_{max})$. If so, equation (6.1) can be re-written:

$$\gamma(w_{max})^{S-1} = w_{max}/w_{min}. \qquad (6.1.1)$$

This equation is readily solved for S to yield an approximate solution
for species richness:

$$S = 1 + \ln(w_{max}/(w_{min}))/\ln(\gamma(w_{max})) \qquad (6.1.2)$$

The term $\gamma(w_{max})$ has the following general form, following equation
(5.16) from chapter 5:

$$\gamma(w_{max}) = [1 + (kw_{max}^{-\delta})^{1/2}]^{1/(F(1+a))}, \qquad (6.1.3)$$

where the limiting similarity constant $k = [B_1(1 + a)(1 + b)]/[\theta^{(1+a)}\sigma^{(1+b)}v]$,
as derived in chapter 5, given $a = (Q - F)/D$, $b = (2F + Q)/D$, $\theta = (1 + a)/(2 + a)$ and $\sigma = (1 + b)/(2 + b)$, as derived in chapter 4. The limiting
similarity exponent $\delta = Q(1 + b) + D + F(a - b) - 9/4$ was also derived
in chapter 5. If we let the variable $y = w_{max}^{-\delta/2}$, then the sum inside the
brackets is the first two terms of a series approximation for $e^{\sqrt{ky}}$:

$$e^{\sqrt{ky}} \cong 1 + \sqrt{ky} + ky^2/2 + \dots. \qquad (6.1.4)$$

Because y is less than 1 for positive δ, the series converges and the
higher order terms are very small and can be ignored, yielding:

$$\gamma(w_{max}) \cong (e^{\sqrt{ky}})^{1/(F(1+a))}. \qquad (6.1.5)$$

Substituting for y into equation (6.1.5) and then for $\gamma(w_{max})$ in equation
(6.1.2) yields:

(Box 6.1 continued)

$$S = 1 + [F(1 + a)/\sqrt{k}]\ln(w_{max}/w_{min})w_{max}^{\delta/2}. \qquad (6.1.6)$$

Substituting in for w_{max} and w_{min} from their derived equations (4.10) and (5.19) in chapters 4 and 5,

$$w_{max} = c_P^{1/(F(1+F/D))} x^{1/(1+F/D)}/\theta^{1/F} \text{ and } w_{min} = (\theta\sigma)^{-1/Q}$$

yields:

$$S = 1 + [F(1 + a)/(\theta^{\delta/2F}\sqrt{k})]\ln(c_P^{1/(F(1+F/D))}x^{1/(1+F/D)}\theta^{(1/Q-1/F)}\sigma^{1/Q})$$
$$\cdot c_P^{\delta/(2F(1+F/D))}x^{\delta/2(1+F/D)}. \qquad (6.1.7)$$

Substituting area for extent, $x = A^{1/2}$, and letting the scaling exponent for maximum food cluster size $\phi = 1/(1 + F/D)$ yields a final

$$S = 1 + [F(1 + a)/(\theta^{\delta/2F}\sqrt{k})]\ln(c_P^{\phi/F}\theta^{(1/Q-1/F)}\sigma^{1/Q}A^{\phi/2})\, c_P^{\delta\phi/2F}A^{\delta\phi/4}. \qquad (6.1.8)$$

Combining constants (variables other than area) yields:

$$S = 1 + k_1 A^{\delta\phi/4}\ln(k_2 A^{\phi/2}), \qquad (6.1.9)$$

where $k_1 = [F(1 + a)/(\theta^{\delta/2F}\sqrt{k})]c_P^{\delta\phi/2F}$ and $k_2 = c_P^{\phi/F}\theta^{(1/Q-1/F)}\sigma^{1/Q}$.

---

power law portion of the solution for species richness becomes the dominant term. A bit of analysis (fig. 6.1) reveals that, at *large* scales (areas) of observation, the predicted relationship may very closely resemble a scaling law. In contrast, under conditions where food is abundant (high F, and thus low ϕ) but mean resource concentrations are low so that Q is small relative to F (leading to low δ), the exponent of the power law portion of the species-area relationship ($\phi\delta/4$) may be so small that the logarithmic portion dominates leading to a predicted species-area curve that will be semi-logarithmic. This formulation may be the first species-area relation predicted by a niche-assembly model.

Because scaling law relationships between species diversity and area are so widely observed, one might be tempted to write off the spatial scaling model as a viable hypothesis of species diversity. However, a closer inspection of data suggests that scaling law relationships may not be so universal (Harte et al. 2005), and that semi-log relationships may be quite consistent with size ratio and abundance vs. size

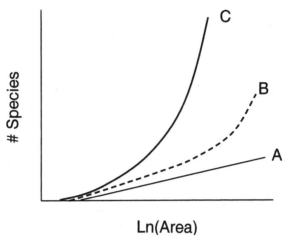

FIGURE 6.1. Hypothetical species-area curves predicted by the spatial scaling model for three different resource scenarios. (A) Rare low-quality food, (B) abundant low-quality or rare high-quality food, and (C) abundant high-quality food. Curve (A) is virtually a semi-log relationship, while (C) is virtually a power law relationship. Curve (B) resembles a semi-log relationship at small areas, but resembles a power law at large scales.

relationships that support the idea that communities are structured by resource partitioning for exclusive resources. The spatial scaling model's complicated, non-scaling law form is similar to Hubbell's (2001) model for dispersal-limited metacommunities and the model of maximum information entropy (Harte et al. 2008), in which individuals of species are assumed to have the distribution that could occur in the maximum number of possible ways. One reason for the lack of power law species-area curves in theory but their prevalence in published empirical studies is that, over 4–5 orders of magnitude in area, semi-logarithmic patterns and power laws may provide similar fits for species-area relationships, especially if the area sampled is much larger than the maximum sampling scale w_{max} in the community.

For three of the seven guilds I explored in chapter 5, dung beetles, grasshoppers, and herbaceous plants, I gathered additional data in samples of sufficiently different areas to construct species-area relationships (fig. 6.2). For these three data sets, I used the same model fitted to their size ratio vs. size relationship (figs. 5.6–5.9) to calculate the expected number of species for different areas A for which I had

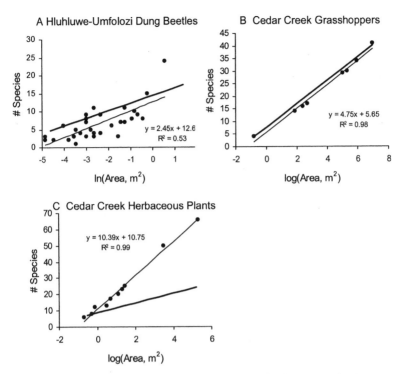

FIGURE 6.2. Species-area relationships for (A) South African dung beetles sampled from dung patches of different area, (B) grasshoppers, and (C) herbaceous plants sampled in nested plots of different area in the same Minnesota old field. Heavy lines are the predicted relationship from the spatial scaling model fitted to size ratio vs. size relationships, thin lines are best-fit semi-log regressions (see chapter 5).

data. The success of the spatial scaling model varied considerably among guilds. The model predicted a slope very similar to that of observed data for dung beetles, but consistently over-predicted the number of species across the range of dung patch areas sampled. The model predicted the species-area relationship for grasshoppers sampled in nested plots within an old field spectacularly well, as the model relationship was virtually identical to the least squares best-fit regression line through the data. In contrast to these two animal guilds, the model predicted the species-area relationship for plants very poorly, as it dramatically under-predicted the number of species found.

These variable results suggest that the spatial scaling model works better for some guilds than others. With only three guilds analyzed, it is difficult to generalize, but the results may reflect the influence of dispersal limitation in determining species richness. Grasshoppers are highly mobile, and their food is space-filling (table 5.1 in chapter 5), so they are unlikely to be limited in their ability to find acceptable food clusters. In contrast, dung beetle food is distributed as widely separated discrete patches and is not space-filling (table 5.1), and it is plausible that some species might be limited in their ability to disperse to all suitable patches, thus reducing species richness. Finally, the species richness of herbaceous plants at the Cedar Creek site is known to be dispersal- and colonization-limited (Tilman 1994), not because resources are rare and patchy but simply because plants are relatively immobile and many species have difficulty dispersing seeds more than a few meters each generation. The strength of the qualitative predictions of spatial scaling for species-area relationships is therefore inversely related to the hypothesized degree of dispersal limitation for the guilds I examined.

These results suggest that, if communities are unlikely to be dispersal-limited, the model may predict species-area relationships very well. In contrast, if species are limited by their ability to disperse to suitable food and resource clusters, such as in dynamics modeled by Hubbell's (2001) neutral theory and other models (Pacala and Tilman 1994; Tilman 1994), then different species-area relationships may result, and the "signal" of dispersal limitation may cause important deviations from the pattern expected from the spatial scaling model. I deal with this issue in more detail in chapter 9. Regardless, the spatial scaling model makes realistic qualitative predictions of the form of species-area relations, sometimes with spectacularly close agreement to data. The model thus serves as a viable alternative hypothesis to explain how species richness increases with area sampled.

Diversity vs. Productivity

The relationship between diversity and productivity has received intense interest over the past 15 years, probably because of its importance in understanding the value of biodiversity and the effects of global changes in atmospheric carbon dioxide and nitrogen on biodiversity (Sala et al. 2000; Reich et al. 2001, 2006; Fukami and Wardle 2005). Recent reviews of multiple relationships between diversity and productivity suggest that many different patterns can result, depending mainly on the scale (Chase and Leibold 2003; Harrison et al. 2006) and type of taxa sampled (Dodson et al. 2000), the interpretation and measurement of "productivity" (Waide et al. 1999; Hawkins et al. 2003; Currie et al 2004), and the importance of predators in regulating abundance and diversity of species (Worm et al. 2002).

The spatial scaling model of community structure can be modified to predict species diversity as a function of productivity. First, we must define how the variables in the model reflect "productivity." A way of calculating productivity that meshes well with previous empirical studies is to calculate the rate of supply of food for a given constant resource concentration. In a fractal environment, this is the resource density x^{Q-D} divided by its renewal time τ. In the spatial scaling model, productivity, independently of observer scale, increases exponentially with the resource fractal dimension Q and is inversely proportional to v, the number of sampling volumes encountered in time τ.

Inspection of equation (6.2) reveals that species richness is proportional to increasing v (imbedded in the constant k) and thus renewal time τ, implying that diversity should decrease with increasing productivity when productivity is manifested by a *more rapid renewal* of resources (smaller τ and thus smaller v). Decreasing renewal time effectively decreases how many volumes can be sampled before resources are renewed and restricts the opportunity for resource partitioning among species with different sampling scales w. At faster resource renewal, regardless of resource distribution, different-scaled consumers will tend to select very similar food clusters and resource concentrations; and larger-scaled consumers lose the relative advantage of being able to sample a larger fraction of the environment. Consumers effectively become "slower" in their ability to reduce resources, and less selective in their food cluster selection. Thus species must be much more different in sampling scale in order to coexist, and more loosely packed in the niche space.

In contrast, the factor δ in the exponents in equation (6.2) is proportional to Q, the fractal dimension of resources. Species richness is therefore positively related to Q and to productivity when it is manifested by

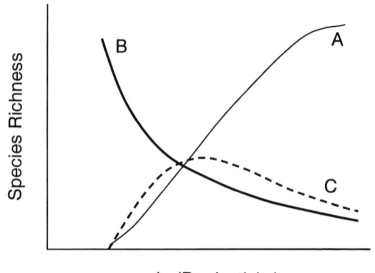

In (Productivity)

FIGURE 6.3. Hypothetical curves of species richness as a function of re-source productivity predicted by the spatial scaling model. (A) Increase in species richness (thin solid line) when higher productivity reflects an in-crease in mass density but not a decrease in the resource renewal interval (proportional to $1/v$). (B) Decline in species richness (thick solid line) when increased productivity reflects no change in mass but a decrease in resource renewal interval. (C) Unimodal change in species richness (dashed line) possible when productivity reflects both increased mass and decreased re-source renewal interval.

greater *space-filling or mass* of resources. A higher Q translates into larger potential resource clusters and greater opportunity for scale-de-pendent resource partitioning and coexistence. Species can coexist on more similar size and resource concentration of the food clusters, and thus be more similar in sampling scale, leading to greater packing of species. This has a direct analogy to the idea that diversity is a function of environmental capacity, such as energy in the form of water or tem-perature (Hawkins et al. 2003; Currie et al. 2004) or limiting elements.

These contrasting qualitative changes in species richness with dif-ferent aspects of productivity produce a rich set of possible diversity-productivity relationships (fig. 6.3), including various increasing, decreasing, or unimodal relationships. The actual relationship obtained will depend on the range of variation in Q and v with different productivities and the scale of observation. For example, if environments do not differ much in the standing mass of resources but differ greatly in their turnover or renewal rate, then species richness should decline with increasing productivity. One scenario where this might occur is in plant systems limited by soil nutrients. In more productive habitats (fertile soils), nutrients would not only appear at higher concentrations, but would likely renew more rapidly, as measured by higher nutrient mineralization rates. If so, we might expect the diversity of organisms, such as nutrient-limited plants, to decline with increasing productivity (Grime 1979; Tilman 1982, 1990; Greig-Smith 1983).

In contrast, if environments differ greatly in the mass of resources supplied but not so much in their renewal rate, then species richness should simply increase linearly or monotonically with productivity (fig. 6.3). One example of such a resource might be radiant energy, measured as light and temperature, which is supplied continuously but at different levels across environments such as from the equator to the poles. Thus we might expect light- or temperature-limited organisms, such as plants, amphibians, or even arthropods to exhibit monotonically increasing diversity in response to increased productivity (Francis and Currie 2003; Hawkins et al. 2003; Currie et al. 2004).

In most environments, higher productivity probably results in both greater mass and faster renewal (shorter renewal times), and the fractal heterogeneity model predicts that the relative magnitude of these changes across a productivity gradient will determine the shape of the diversity-productivity relationship. One interesting qualitative prediction is that the influence of resource renewal rate relative to resource mass should be relatively stronger at smaller scales of observation. This can be readily seen in equation (6.2) because the influence of mass is reflected in the factor δ in the exponent of A or area observed, whereas the influence of v (through the parameter k) is independent of area observed. As area increases, the influence of mass on species richness increases, making positive diversity-productivity relationships more likely. This means that negative or unimodal diversity-productivity relationships might be more prevalent at "local" scales, while positive relationships might be more prevalent at larger scales (fig. 6.4A). This qualitative prediction of observer scale-dependent

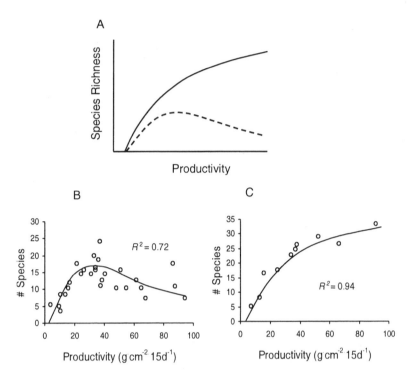

FIGURE 6.4. (A) Hypothetical shift in the diversity-productivity relationship from unimodal at small (dashed curve) to monotonic at large (thin curve) scales of observation (areas sampled) predicted by the fractal heterogeneity model. (B) Species richness of pond animals in local temperate freshwater ponds of different productivity, fit to the spatial scaling model with adjustable parameters v and Q ($R^2 = 0.72$, $P < 0.0001$). (C) Watershed-level species richness, aggregated across 3 ponds in each of 10 different watersheds whose ponds differ in mean productivity, fit to the spatial scaling model with adjustable v and Q ($R^2 = 0.94$, $P < 0.0001$). Data from Chase and Leibold (2002). Note that in both cases the spatial scaling model fit the data better than the quadratic ($R^2 = 0.33$) and linear ($R^2 = 0.74$) functions fit to (A) and (B), respectively, by Chase and Leibold (2002).

diversity-productivity relationships matches that observed in many literature reviews and recent studies of plants (Gross et al. 2000; Rajaniemi 2003; Venterink et al. 2003; Harrison et al. 2006), pond animals (Chase and Leibold 2003; Chase and Ryberg 2004) and pelagic lake fauna (Dodson et al. 2000). In particular Chase and Leibold's (2002) study of aquatic plants and animals in ponds with different productivity beautifully supports the prediction (fig. 6.4B,C).

The scale-dependence can be viewed as an interaction between scale of observation and productivity: In unproductive environments, increasing area has relatively little impact on the variance in food cluster size (all are small even in large landscapes) and provides little additional opportunity for resource partitioning. This means that the difference between the diversity-productivity curves for different areas sampled will be small at low productivity (fig. 6.3). In productive environments, however, the range of available food cluster sizes and resource concentrations increases as the exponent Q and thus dramatically with area, and may allow sufficiently greater opportunities for resource partitioning to overcome the negative effect of decreasing renewal time at higher productivity. Thus the difference in species richness between samples from different areas will be much greater at higher productivity (fig. 6.3). Chase and Leibold (2002) and later Chase and Ryberg (2004) show convincingly that the scale-dependence in diversity-productivity relationship results from greater beta diversity and thus species turnover across space at higher productivity. Such turnover necessarily would occur as the result of resource partitioning and species selection of different food cluster sizes across a landscape, as argued by the spatial scaling model. Dispersal limitation may also contribute strongly to this scale dependence (Chase and Ryberg 2004).

I examined diversity-productivity relationships for 3 of the 7 guilds I introduced in chapter 5: Serengeti herbivorous mammals and Cedar Creek, Minnesota plants and grasshoppers. Serengeti mammal species richness was obtained for 1 km² transects during the dry season at sites in Serengeti National Park (Ritchie unpublished data) that differed in their annual productivity, which was driven by differences in mean annual rainfall (400–1300 mm/yr). Grasshopper and plant diversity were sampled (as detailed in chapter 5) across a gradient of measured aboveground biomass production and soil nitrogen availability, respectively (Ritchie 2000). These examples (fig. 6.5A,B,C) support the general qualitative predictions of the spatial scaling model of community structure: that either negative or unimodal diversity-productivity patterns can result from the resource partitioning mechanisms that seem to account well for their community structure. However, as much as I

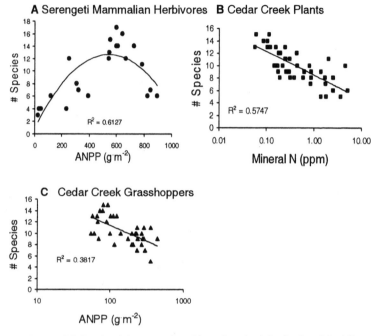

FIGURE 6.5. Species richness vs. logarithm of productivity for 3 guilds. (A) Mammalian herbivore species encountered on 5 km x 200 m transects (Ritchie unpublished data) at locations with different measured ungrazed aboveground net primary productivity (ANPP) in Serengeti National Park, Tanzania (McNaughton 1985, Ruess and Seagle 1995, unpublished data). (B) Herbaceous plants and, (C) grasshoppers sampled from twelve 9 x 9 m plots, half unfertilized and half fertilized with 16 $gm^{-2}yr^{-1}$ N as NH_4NO_3, in which we measured available soil mineral N concentrations (Ritchie 2000).

would like to, it is not feasible to make specific quantitative predictions of diversity-productivity relationships for these guilds from the parameters fitted to their size ratio data, simply because it is unknown how the renewal time for resources for these guilds changed with productivity.

The principle conclusion from the analysis of diversity-productivity relationships is that: (1) ecologists need to account for changes in renewal time as well as in mass produced in measuring productivity; (2) many different diversity-productivity relationships are possible at a given scale of observation, and the incredibly variable set of patterns observed in literature reviews (Waide et al. 1999; Mittelbach 2001) is exactly what you might expect, and confirms the lack of any general or universal pattern; and (3) patterns for environments sampled at larger spatial scales relative to the size of the organisms of interest are more likely to be positive and monotonic than environments sampled at smaller or "local" scales. The probability statement in (3) implies that sampling at a larger scale does not guarantee a positive relationship; negative, unimodal, or neutral relationships can still occur if the decrease in resource renewal time with increasing productivity is strong enough. These results and interpretations suggest some new directions for research on the influence of productivity on diversity, and attest to the futility of searching for universal diversity-productivity relationships or for inferring mechanisms from patterns (Mittelbach et al. 2001).

Diversity and Abundance in Heterogeneous Environments

Another major class of diversity patterns mixes species diversity with abundance. Two favorites are the species-abundance distribution or so-called "log-rank abundance curve" (MacArthur 1972; Whittaker 1975; Hubbell 2001) and the diversity-abundance distribution (Preston 1962; Hubbell 2001; McGill 2003; McGill et al. 2006). Neither of these patterns has been used to derive niche-assembly models of species diversity, but I suspect it is because no one has bothered to do it, rather than that it cannot be done. For example, the energy equivalence rule (EER) (Damuth 1981; Brown 1995) implies species-abundance and diversity-abundance relationships that can be compared with observed data and the quantitative predictions of other models, but no studies of which I am aware have explored this.

The spatial scaling model predicts an equilibrium abundance for each species allowed by the species-packing rule, so it is straightforward to calculate species- abundance distributions (log-rank abundance curves) for the spatial scaling model of community structure (fig. 6.6). The spatial scaling model predicts species-abundance relationships that are visually distinct from neutral models determined from the total abundance and species richness from the same simulation (fig. 6.6). The spatial scaling model predicts that the most abundant species will

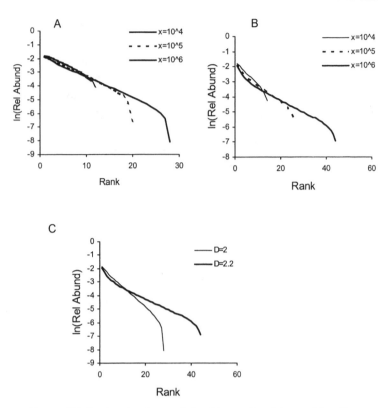

FIGURE 6.6. Species-abundance relationships predicted the spatial scaling model for (A) three different scales of observation ($x = 10^4$, 10^5, 10^6) that reflect a 10,000-fold increase in area sampled for consumers in a 2-dimensional environment, (B) for consumers in an environment with an additional vertical dimension ($D = 2.2$), and (C) for consumers sampled at a single scale for $D = 2$ and $D = 2.2$. For these simulations, $F = 1.7$, $Q = 1.4$, $v = 10^4$, and $B_1 = 30$. Note the emergence of a lognormal-appearing species-abundance distribution and a greater coexistence of rare species with the inclusion of a vertical dimension.

be relatively more dominant than expected from the lognormal at small scales of observation, and that rare species will be relatively more dominant than expected from the logseries at large scales of observation. Species-abundance distributions are most sensitive to the environmental dimension D, which can exceed 2 if there is a vertical dimension to the habitat. A greater vertical dimension, or increasing dimensionality in general, tends to curve the relative abundance-rank relationship such that the most abundant species become relatively less dominant and many more rare species can coexist (fig. 6.6C). For a given environmental dimension, increasing the scale of observation (fig. 6.6A) has relatively little effect on the abundance-rank relationship, as does increasing food or resource density (increasing F or Q). Where $D > 2$, increasing the scale of observation shifts the rank-abundance curve from a more linear form at small scales to a more inverted S-shape reminiscent of the curve from a lognormal distribution at larger scales. This scale and dimensional dependence of the form of the species abundance curve is implicitly predicted in some neutral models, such as by increasing Hubbell's (2001) biodiversity number, but has not to my knowledge been predicted from any niche assembly model.

The predicted species abundance distributions can be compared with observed data and also alternative "neutral" statistically derived models, such as the lognormal, geometric series, and logseries (Preston 1962; May 1975; Whittaker 1975; Magurran 2004) that do not depend on any niche assembly processes. A lognormal species abundance distribution is thought to result from either a large number of species of independent population dynamics with randomly varying (in either space or time) exponential growth, such that $N(i) \propto e^{r_i}$ where r_i is a random variable. Since $N(i)$ is a function of an exponential variable, $\ln(N(i))$ should be normally distributed (May 1975). Alternatively, species in a community that are limited by multiple factors that act on population size in a multiplicative fashion should also exhibit a lognormal distribution of abundances. I generated the lognormal species-abundance distribution for each sample with the observed mean ln(abundance) and I estimated standard deviation of the abundance distribution as the range divided by 4, so as to always include a species abundance of 1 in the distribution. These values were used in the normal distribution macro program in Excel® to generate three different abundance distributions for the S species in each sample, which were then averaged to obtain a set of ranked abundances.

A geometric series distribution is thought to derive from priority exploitation of resources by species arriving sequentially in a community (Whittaker 1975). An expected geometric series is obtained by assuming that each species' abundance is proportional to a fixed proportion p of remaining resources. Thus the relative abundance of the ith species is $(1 - p)p^{i-1}$. I used the relative abundance of the most abundant species to estimate p for calculating the expected geometric series' species-abundance curve.

Another statistical distribution is the logseries, which can result from random dispersal from some larger community, as in Hubbell (2001), or other mechanisms. In fact, the logseries is the asymptotic species abundance distribution for the analytical version of Hubbell's neutral model (Volkov et al. 2003) and Harte's maximum entropy-derived state variable model (Harte et al. 2008). For a logseries distribution, species having abundance n occur with frequency $\alpha x^n/n$, where x is a fitted parameter and α is Fisher's alpha, a measure of species diversity that is independent of total community abundance. For a given community with N total individuals and S species, x can be found (Magurran 2004) by iteratively solving the following equation for x: $S/N = -\ln(1 - x)(1 - x)/x$ and then finding Fisher's alpha as $\alpha = N(1 - x)/x$. Then x and α are used to calculate the frequency distribution of n and thus a species-abundance curve for a given N and S. I used S and N from each guild to iteratively estimate x and calculate Fisher's α and the expected logseries species-abundance curve.

I compared observed species-abundance distributions with the predicted ones for the spatial scaling, lognormal, geometric and logseries distributions for four guilds, mammalian herbivores, dung beetles, and grasshoppers. Comparisons were made with an Akaike Goodness of fit test (Burnham and Anderson 2002). In this test, an Akaike Information Criterion (AIC) was determined as the natural logarithm of the mean (sum divided by S) of squared deviations between observed and predicted ln(relative abundance) for all ranked S species plus an additional term to correct for the number of estimated parameters, k (1 for geometric series and 2 each for logseries and lognormal distributions): $(S + k)/(S - k - 2)$. The lower the calculated AIC value, the better the fit. A difference of 1 in AIC corresponds roughly to a 3-fold difference in sums of squared deviations. The lognormal and logseries models each used two parameters ($k = 2$) estimated from the data, while the geometric used one ($k = 1$). Although the spatial scaling model has 6 potentially adjustable parameters, in reality only 3 were adjusted to fit the size distribution data for each guild, and the model used no data on species richness or abundance to predict the species abundance distribution, so $k = 0$.

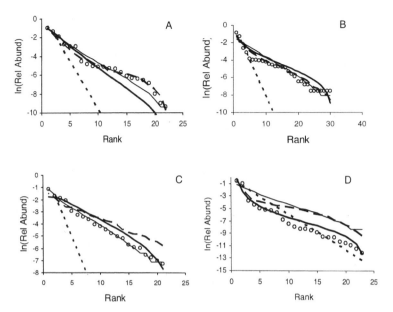

FIGURE 6.7. Species-abundance relationships for four guilds: (A) Serengeti (Tanzania) mammalian herbivores, (B) Hluhluwe-Umfolozi (South Africa) dung beetles, (C) Cedar Creek (Minnesota) grasshoppers, and (D) Cedar Creek plants. For species-abundance relationships, circles represent observed ln(relative abundance) and the lines are the predictions from the lognormal (gray heavy dashed line), spatial scaling (heavy line), geometric series (thin dashed line), and logseries (thin line) distributions. The fractal model was calculated from the model fit to size ratio vs. size relationships for each guild in chapter 5. Distributions for lognormal, geometric, and logseries were calculated as discussed in the text.

The observed species abundance distributions were fairly well predicted by the spatial scaling model, particularly for the most abundant third to half of species (fig. 6.7). However, the spatial scaling model more poorly predicted the relative abundance of the rarest third to half of species. The spatial scaling model always outperformed (lower AIC) the geometric series, and outperformed all of the other statistical distributions for plants. It was outperformed by the lognormal model for Serengeti herbivores and dung beetles and by the logseries for grasshoppers, largely because of its weakness in predicting the relative abundance of rare species. The ability to predict relative abundances of common species but not rare ones may reflect inherent differences in the mechanisms that determine community membership. For example, Magurran and Henderson (2003) suggest that most rare species are transients that would otherwise not be able to permanently persist in the presence of core species. This study and a recent survey of tintinnids from the Mediterranean (Dolan et al. 2008) suggest that core and transient species differ in their species abundance distributions and that this is likely the structuring mechanism for their occurrence in communities.

This analysis highlights the potential for a niche-assembly mechanism, that is, the competitive coexistence of species of different size on similar but fractally distributed resources, to provide an explicit quantitative prediction of species abundance distributions. It predicts the abundance of dominant species much better than neutral statistical models. The relatively large deviation in the abundances of rare species may again, as we observed for species-area curves, reflect the "signal" of dispersal limitation or other community assembly mechanisms.

Local vs. Regional Species Richness

A considerable literature has developed around the premise that communities structured by competition or other species interactions should have limited local membership and a finite, saturated species richness, regardless of the number of species in the much larger "regional" area in which the local environment is embedded (Cornell and Lawton 1992; Cornell 1999; Lawton 1999). In contrast, communities weakly structured by species interactions but whose membership is strongly dispersal-limited should have unsaturated local communities whose species richness increases in proportion to regional richness. These different expected relationships (summarized in fig. 6.8A) have been used extensively as diagnostics of interaction-structured vs. unsaturated dispersal-structured communities.

Despite the volume of this literature, recent studies of communities of zooplankton (Shurin and Allen 2001; Havel and Shurin 2004) and plants (Tilman 1994) which are known to be heavily structured by competition and predation interactions, show a proportional relationship between local and regional diversity. These recent data call to

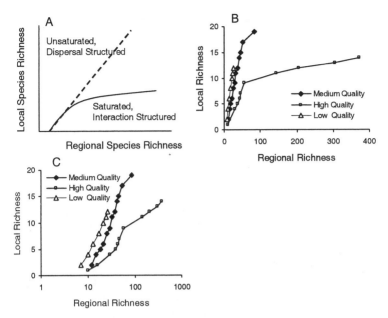

FIGURE 6.8. Theoretical relationships between local species richness and regional species richness. (A) Hypothetical local–regional richness relationships predicted for local communities structured by species interactions (thin curve) vs. structured by dispersal limitation (dashed line), as suggested by Cornell and Lawton (1992). (B) and (C) Simulations of local and regional richness for fully saturated coexistence communities, as predicted by the spatial scaling model, for guilds inhabiting local and regional environments that differ by 10^6 in area, with species constant $v = 10^4$, $D = 2$, and $B_1 = 40$. Food and resource availability were varied to generate different regional richness values for 3 different resource scenarios: medium (solid diamonds), high (open squares), and low (open triangles) resource quality.

question the indiscriminant use of local vs. regional richness plots as diagnostics of community structure. However, ecologists still don't understand fully why local and regional richness should be linearly correlated if local communities are structured by species interactions.

The spatial scaling model of community structure offers, for the first time, the opportunity to compare "local" and "regional" species richness of communities that are explicitly structured by competition. I simulated such curves for 3 different resource scenarios where variation in local ($x = 10^4$) and regional ($x = 10^7$) species richness was driven by differences in food quality and mobility (v); given that resource requirements (B_1) and environmental dimension (D) were held constant (fig. 6.8B,C): low quality food, $F - Q = 0.4$, medium quality food, $F - Q = 0.3$, and high quality food, $F - Q = 0.2$. These simulations revealed no saturation in local richness with increasing regional species richness for low and medium quality resources, which correspond roughly to food with <5% resource. For high quality food (5–10% resource), I found saturation of local richness but only in environments with high food abundance ($F > 1.6$) where regional richness increased dramatically with an increase in the scale of observation (x or A).

These results reflect a general property of the spatial scaling model where, despite the assumption of no dispersal limitation, local and regional richness are roughly linearly correlated across the full range of possible regional species richness, except under a relatively narrow range of conditions. For any given set of species of unknown resource distribution, it would be highly probable that local and regional richness would be linearly correlated even if they strongly compete for a limiting resource and experience no dispersal limitation. Therefore, evaluation of linearity or saturation of plots of local and regional richness are likely virtually useless in detecting the signal of dispersal limitation and the strength of species interactions in structuring communities.

The linear correlation between local and regional richness predicted by the spatial scaling model illustrates a major point that is often forgotten in the discussion of local and regional richness: regional richness is itself composed of local richness. Any process that structures local richness likely also structures the pool of potential species in the regional pool. Assembly of large-scale communities thus reflects "bottom-up" accumulation of structuring of local communities as much as it reflects the fact that local communities are constrained in their membership by the regional species pool. While factors other than local structuring mechanisms may also influence regional richness, the cross-scale signal of local structure will likely lead to some linearity in the correlation of local and regional richness (Shurin and Allen 2001; Havel and Shurin 2004).

These results do not suggest that local communities cannot be dispersal limited. The analysis merely suggests that a significantly positive slope of a linear relationship between local and regional richness is not sufficient to determine dispersal limitation. The spatial scaling model instead offers a potentially powerful method for evaluating the strength of dispersal limitation in structuring local communities, as it predicts the relationship between regional and local species richness under the assumption of no dispersal limitation. If species cannot always disperse to food and resource clusters they might exclusively use, fewer species than expected from the spatial scaling model should be detected for a given observed regional richness. This seems likely for most data sets, as for the wide variety of food abundances and resource concentrations explored in the low and medium food quality simulations, maximum regional richness for a region one million times the size of the local community was less than 50. This should show up in a steeper than expected species-area curve as well. This quantitative prediction is supported earlier in this chapter where Cedar Creek plants, which are known to be dispersal limited (Tilman 1994), have a much steeper species-area curve than predicted by the spatial scaling model. Thus, the explicit quantitative prediction of local vs. regional species richness relationships made by the spatial scaling model provides a theoretical "fulcrum" for detecting the magnitude of dispersal and resource competition in the community structure of guilds.

SUMMARY

1. Existing niche assembly models of species diversity predict relatively few species to coexist and have no explicit way of incorporating spatial variation. Few, if any models have explicitly predicted major community patterns, such as species-area, diversity-productivity, and species-abundance relationships, and offer no current alternatives to predictions of neutral theory.

2. Analytical equations for species richness of guilds can be derived from the spatial scaling model and used to predict major diversity patterns.

3. The spatial scaling model predicts a complex relationship between species richness and area that is not a universal power law; species richness is related to ln(area) at low productivity or food resource concentrations and small area sampled, but exhibits a relationship

resembling a power law at high productivity or food resource concentrations and large area sampled.

4. Quantitative predictions of observed species-area relationships for three guilds (dung beetles, grasshoppers, and plants) varied in their success, with accuracy varying inversely proportionally to the suspected degree of dispersal limitation of the different guilds.

5. The spatial scaling model predicts varying and non-universal relationships between species richness and productivity in which richness is positively related to the mass of resources produced, but negatively related to the renewal time of resources. The interplay of these two contrasting effects produces a unimodal diversity-productivity relationship under many conditions. Increasing the scale of observation increases the relative sensitivity of species richness to the mass of resources produced, such that diversity-productivity relationships are very often positive and monotonic at large scales of observation and negative or unimodal at small scales.

6. The spatial scaling model predicts species abundance relationships that fit the relative abundance of the most dominant third of species for four different guilds (mammalian herbivores, dung beetles, grasshoppers, plants), but poorly fit the relative abundance of the rarest third of species. The species abundance distribution for communities from large areas or environments with a significant vertical dimension strongly resembles the species abundance distribution expected from a neutral lognormal model, which implies that the neutrality of a community cannot be judged from its species abundance distribution.

7. The spatial scaling model predicts linear relationships between local and regional species richness for a wide range of conditions, in contrast to the monotonic relationship expected from a plethora of literature studies. This result suggests that linearity of the local vs. regional species richness relationship does not necessarily imply dispersal limitation and a lack of saturation of communities. Rather the spatial scaling model provides an explicit quantitative prediction of local vs. regional richness from a non-dispersal limited, niche assembly mechanism that can be compared with observed data to detect the degree of community saturation and dispersal limitation. The spatial scaling model provides a comprehensive niche-assembly-based theory of community structure and species

diversity that, when its assumptions of resource rather than dispersal limitation are met, predicts observed data very closely. The model thus offers for the first time a viable niche-assembly hypothesis for the structure of communities that can be compared with the predictions of other theory.

Biodiversity Conservation in Fractal Landscapes

The spatial scaling model of consumer-resource interactions yields qualitative predictions about the richness and body size, or other morphological traits of species, in environments that differ in the amount and spatial pattern of resources. Consequently the spatial scaling model can be used to address some questions relevant to the conservation of biodiversity. One consequence of environmental changes on communities that can be uniquely addressed with this model is the effect of habitat fragmentation on species' populations and community species richness for particular guilds that use the same resources in the same habitats.

A major question faced by conservationists is how much habitat, such as tropical rainforest, freshwater bodies, grassland, desert, etc., can be lost; and for a given habitat amount, to what degree can it be broken up or *fragmented* into smaller pieces before a species goes extinct (Andrén 1994; With and Crist 1995; With and King 1999; Fahrig 2002) or species are lost (Palmer 1992; Rosenzweig 1995). This is a central question in the design of reserves and alternative land uses for natural ecosystems, and in assessing the effects of natural and anthropogenic disturbances on populations and communities of species.

Habitat can be described as the space in which a species can tolerate the extant physical conditions, biotic influences on mortality, and in which their limiting resources occur in sufficient abundance. The habitat is assumed to contain, and thus have nested within it, the food and resources as described in previous chapters. To explore how variation in the amount and distribution of habitat might affect species' persistence in a landscape and thus species richness, I illustrate how the dimension H of a fractal habitat reflects information about density, dispersion, and fragmentation. I then explore how changes in H, corresponding to habitat loss or aggregation, might contribute to species loss and changes in size distributions. This involves changes in habitat area as well as the degree to which a given habitat area is "fragmented" or dispersed into many small vs. fewer, larger clusters or "patches."

HABITAT LOSS, FRAGMENTATION, AND FRACTAL GEOMETRY

Fractal geometry provides a tool for quantifying habitat fragmentation that is sensitive to the scale at which a habitat distribution is viewed (*extent*), and the resolution ε at which space is classified as habitat vs. non-habitat (With and King 1999). Ideally this would be the same resolution at which resource and food distributions are classified (typically $\varepsilon = 1 - 10$ mm), but in practice, habitat is likely to be classified at a much coarser scale ($\varepsilon > 1$ m). However, across large landscapes ($x > 10^7$), it is reasonable to assume that all available food and resources occur within space classified as habitat, and thus, that food and resources are *nested* inside habitat. Mathematically this means that the fractal dimensions of resources and food must be equal to or less than that of habitat: $Q \leq F \leq H$. Consequently, organisms of interest are assumed to occupy, strongly prefer, or can only achieve breeding status in a certain habitat, represented as a binary map of occupied (black) or unoccupied (white) pixels of resolution ε on a landscape of extent x (fig. 7.1). If it is also assumed that the habitat is distributed as a fractal, its features are easily revealed by a calculation of the mass fractal dimension H, that measures the slope of the relationships between ln(occupied pixels) and ln(length) of a window centered on an occupied pixel (Milne 1992) or other appropriate measures (Johnson et al. 1996; Olff and Ritchie 2002; Halley et al. 2004). This habitat will occupy a proportion h of the landscape observed and fill space according to a fractal dimension H. If this habitat is uniformly or randomly distributed and occupies most of the landscape, then H will approach the value of D, and the issue of fragmentation will be unimportant. In contrast, as the amount of habitat declines, or as remaining habitat becomes increasingly clustered by the process of fragmentation, H will decline.

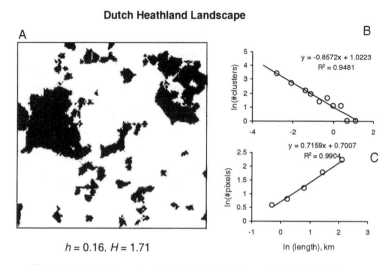

$h = 0.16,\ H = 1.71$

FIGURE 7.1. (A) Example of a fractal distribution of habitat (black) on a landscape, showing the actual distribution of heathland habitat in a 9 x 9 km grid in the Netherlands (Olff and Ritchie 2002). The fractal dimension of habitat in a 2-dimensional landscape (D = 2) can be detected by either, (B) calculating the Korcak exponent $-H/2$, or -2 times the slope of the logarithm of the number of habitat clusters, or "patches" vs. the logarithm of the length of each cluster (see chapter 2) or, (C) calculating the mass fractal H, from a series of line transects across the image, where $H = 1+$ slope of the logarithm of number of units of resolution ε detected in intervals of different length surrounding each occupied interval of the transect (Voss 1986).

The separate influence of clustering, or aggregation of habitat on the fractal dimension H is evident in a comparison of 2 landscapes with similar proportions of habitat but different frequencies of clusters (fig. 7.2). In the example, based on real distributions of heathland habitat in the Netherlands, the more fragmented habitat has a higher value of H and a more restricted range of habitat cluster sizes, as predicted by the Korcak relationship for the frequency of different cluster sizes in a fractal distribution (see chapter 2). In this case, the number n of habitat clusters or patches greater than a given length w, declines as a power law with increasing w, as is the case for any fractal distribution. More fragmented habitat distributions have a more rapid decline in habitat cluster size with increasing w, and therefore a lower likelihood of the occurrence of a large habitat cluster, and thus food or resource

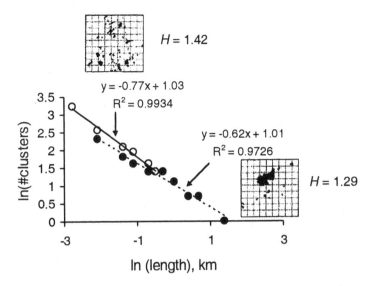

FIGURE 7.2. The logarithm of the number of habitat clusters (sets of contiguously occupied units of resolution ε along a series of line transects), for 2 Dutch landscapes with similar proportions (0.16) of heathland habitat, but different degrees of aggregation, fragmented (open circles) and more aggregated (filled circles). The slopes of the lines are expected to be $-H/2$, where H is the habitat fractal dimension, and thus -0.71 for the fragmented and -0.645 for the aggregated landscape. The slope of the line for the fragmented habitat is significantly steeper ($P < 0.01$) than that for the aggregated habitat. Note also that the frequency of large habitat patches (>1 km) is much higher for the aggregated landscape.

cluster. The more general relationship between increasing amount of habitat h and habitat fractal dimension H, is illustrated by Dutch heathland landscapes (fig. 7.3). This pattern is generally true for any statistically random fractal (Hastings and Sugihara 1993). Overall, increasing h leads to higher H, but for a given h, H declines as habitat is more strongly clustered. This allows, as Han Olff and I pointed out previously (Olff and Ritchie 2002), the effects of amount and distribution (fragmentation) of habitat on species occurrence or diversity to be assessed separately (see also With and King 2004).

The enterprising reader will note that these statements are completely opposite of the relationship between habitat aggregation and fractal dimension reported in Olff and Ritchie (2002). This discrepancy

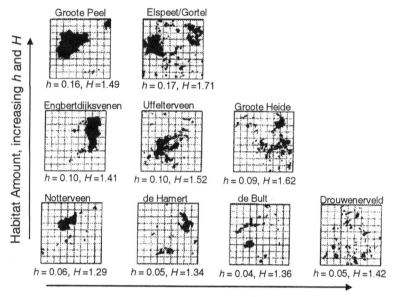

FIGURE 7.3. Example 9 x 9 km fractal landscapes in the Netherlands selected to illustrate the relationship between fractal dimension H of heathland habitat (black) and proportion of landscape occupied by habitat h. Note that H increases with h (bottom row to top row) but also with the degree of habitat aggregation (right to left). Thus for a given amount of habitat h, a lower fractal dimension H indicates a greater degree of fragmentation of the habitat into aggregated clusters.

highlights the importance of identifying the appropriate range of scales over which to evaluate fractal dimension (Halley et al. 2004). In our previous paper (Olff and Ritchie 2002), we used an algorithm that calculated fractal dimension using the mass fractal method (Milne 1992) based on the smallest resolution of the scanned image of published maps of heathland habitat (assuming $\varepsilon = 10$ m). As we pointed out in the paper these maps did not truly differentiate habitat at the resolution we used in our estimation of H. In effect, the maps show fractal distributions that had a "generator," in the sense of the random Cantor carpets discussed in chapter 2, with a much larger actual resolution ε than what we assumed in our calculation. As a result our calculated values of H were equivalent to estimating the fractal dimension *within* the occupied cells of the generator of the fractal, effectively finding that a filled cell in the generator has a dimension of 2. This method therefore overestimated the rate at which habitat area accumulates with increasing scale of observation, and caused us to find that greater aggregation corresponded to a greater fractal dimension. The re-estimated values of H (fig. 7.3) were obtained by using a resolution $\varepsilon = 250$ m, which better reflects the resolution at which the original habitat maps were drawn. The appropriateness of this re-calculation is further borne out by the alternative method of calculating fractal dimension from the frequency distribution of habitat cluster sizes (fig. 7.2).

The theoretical consequences of this for species occupancy and species richness require some further mathematical development. What is needed is an index of fragmentation Φ, that denotes the ratio of the number of habitat clusters Γ to the area of habitat, reflected by the proportional area h multiplied by the area observed x^2, which I have assumed, by convention for the issue of habitat fragmentation and conservation, to be 2-dimensional. Thus,

$$\Phi = \Gamma/hx^2. \tag{7.1}$$

The number of clusters is given by the sum or integral of the number of clusters of different lengths, which is just the frequency given by the Korcak power relationship multiplied by n, the number of clusters with exactly length scale $w = 1$.

$$\Gamma = \int_1^{w_{max}} nw^{-H/D}dw \tag{7.2}$$

Solving the integral yields

$$\Gamma = n[w_{max}^{1-H/D} - 1]/(1 - H/D). \tag{7.3}$$

The next challenge is to determine n given a known h and H and develop an expression for w_{max}. The value of n is constrained by the fact that the sum or integral of mass (or area of landscape covered) of all clusters must sum to the total area of habitat, which of course is hx^D. The landscape area occupied by habitat associated with *each* cluster length is just the mean for the minimum number of windows of length w required to "cover" habitat-occupied spaces in the landscape w^H. Multiplying this cluster size by the frequency yields

$$hx^D = \int_1^{w_{max}} nw^H w^{-H/D} dw. \tag{7.4}$$

Evaluating this integral and rearranging to solve for n gives

$$n = hx^D[w_{max}^{-[1+H(1-1/D)]} - 1]/[1 + H(1 - 1/D)]. \tag{7.5}$$

Distributing food and resources within a restricted set of space influences the chance that a species can persist by affecting the maximum size of food cluster that is likely to be found within the habitat-occupied space. This will affect the maximum required food cluster size P^*_{max} that will occur on the landscape. The original equation (5.19) for w_{max} can be used to qualitatively predict, for a given habitat distribution, the maximum forager sampling scale w that can persist.

$$w_{max} = c_P^{1/(F(1+F/D))} x^{1/(1+F/D)}/\theta^{1/F}, \tag{7.6}$$

where θ and c_P are constants and F is the fractal dimension of food in the landscape of extent x. In this case, the dimension F is an indicator of the distribution of habitat, since habitat is defined to be the space that contains resources. Thus one can argue that $F \leq H$ and

$$w_{max} = k_h x^{1/(1+H/D)}, \tag{7.7}$$

where $k_h = c_P^{1/(H(1+H/D))}/\theta^{1/H}$. Because c_P is an arbitrary small constant (see chapter 2) and $\theta \approx \frac{1}{2}$ (see chapter 4), w_{max} is primarily a function of extent and the amount and distribution of habitat. Substituting w_{max} into equation (7.5) and simplifying yields

$$n = h(1 + H/2)x^{D-1}. \tag{7.8}$$

Assuming that $D = 2$, and substituting equation (7.5) back into equation (7.3), and then substituting (7.3) into equation (7.1) yields a formula for an index of fragmentation Φ:

$$\Phi = (1 + H/2)x^{-2(1+H)/(2+H)}/(1 - H/2). \tag{7.9}$$

This is a very interesting result for two reasons. First, it does not contain h, and thus measures fragmentation independently of habitat area. Second, it shows that fragmentation (interpreted as the number of discrete habitat clusters/habitat area) changes in a complex way with the habitat fractal dimension H (fig. 7.4). From a single point on a land-

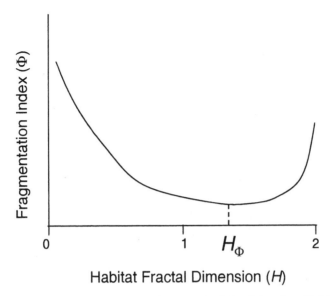

FIGURE 7.4. Relationship between fragmentation index Φ (number of habitat clusters or patches/habitat area) and the habitat fractal dimension H, independent of h, the proportion of landscape occupied by the habitat. H_Φ is the habitat fractal dimension at which fragmentation is minimized or aggregation is maximized for a given landscape extent x.

scape $H = 0$, the fragmentation index declines at a decreasing rate with increasing H, likely in response to the fact that H is most sensitive to the total amount of habitat and not to how it is clustered when the habitat is relatively rare (<3% of the landscape). However, when habitat cover is more abundant than 3%, the fragmentation index Φ, increases at an increasing rate with *greater* habitat fractal dimension H. Simple calculus reveals that the minimum Φ for 2-dimensional landscapes $D = 2$, occurs where

$$H_\Phi = 2[(\ln(x) - 2)/(\ln(x) + 2)]. \tag{7.10}$$

Thus the habitat fractal dimension that corresponds to the greatest habitat aggregation or lowest fragmentation depends only on x, the landscape extent or scale of observation. Small landscapes, like the Dutch heathlands presented in figure 7.3 all have $h \geq 0.04$, and $H > 1.23$, the calculated minimum for a 9 x 9 km, 1 m resolution landscape.

Thus, increasing H for abundant habitats will lead to a greater frequency of small habitat clusters as space becomes increasingly filled, and thus greater fragmentation. Habitats rarer than 1% of a landscape are likely to be irrelevant for conservation analysis, so the patterns of increasing Φ with increasing H are likely true for most situations of conservation interest.

Over the range $H_\Phi \leq H \leq 2$, the fragmentation index can be approximated as

$$\Phi \approx (2 + H)/x(2 - H), \tag{7.11}$$

which is a very simple expression for how fragmentation changes with fractal dimension for abundant habitats. The extent term in the equation partially accounts for the expected proportion of habitat occupied for a given value of H, where a greater x implies a lower h since more extensive landscapes will be occupied by a lower proportion of habitat for a given value of H (see chapter 2). Thus the fragmentation index is inversely related to x and positively related to h. This reflects the lower chance of clustering of habitat at lower overall density of habitat detected at larger scales of observation or in landscapes of greater extent.

SPECIES PERSISTENCE AND HABITAT LOSS AND FRAGMENTATION

With the issue of habitat amount and fragmentation clarified, I now determine the values of these that can allow a species to persist, by solving for the critical H_{crit} that would provide a sufficiently large food cluster size that would support a species of size w in equation (7.7).

$$H_{crit} = \ln(A)/\ln(w/k_h) - 2 \tag{7.12}$$

Because w appears in the denominator of this expression, larger-scaled species require a smaller H and thus greater aggregation of habitat to persist than smaller species. Substituting equation (7.12) into equation (7.11) and remembering that $x = A^{1/2}$, yields the criterion for the fragmentation index explicitly:

$$\Phi_{crit} \leq [\ln(A)/\{A^{1/2}[4 \ln(w/k_h) - \ln(A)]\}]. \tag{7.13}$$

Thus we might expect larger-scaled species with larger w to drop out of communities as H and Φ increase. Ironically, the critical fragmentation increases as the landscape extent (A) increases because of the decline in habitat density that results from expanding the scale of observation or observing larger landscapes. One consequence of this is that the probability of a species' occurrence should decline with increasing H.

Another question can be asked—How large of an area A_{crit} would need to be sampled to find a species of a given sampling scale w, in a habitat with proportional cover h and dimension H (or fragmentation Φ)? This should be roughly equivalent to a species' measured scale or area of response; changes in habitat density at scales smaller or larger than $A^{1/2}$ would likely be responded to dramatically less by a given species. At larger scales, the species can select particular concentrations of habitat within "windows of area A," while at smaller scales, the species might be integrating over variations detected at smaller scales or areas. Solving for A in equation (7.12) yields

$$A = (w/k_h)^{2+H} \tag{7.14}$$

This very interesting result predicts that the scale of response for a consumer species will increase directly with sampling scale w, with a scaling exponent equal to $2 + H$. For expected values of H between 1.2 and 2, this translates to a potential scaling exponent of 3.2 to 4.

SPECIES DIVERSITY AND HABITAT LOSS AND FRAGMENTATION

The relationship between maximum cluster size P_{max}, maximum scale w_{max}, and area (x^2 or A) implies that the influence of habitat distribution on species diversity can be derived from the species-area relationship developed in chapter 6 (eqn. (6.4), and box 6.1).

$$S = 1 + k_1 A^{\delta\phi/4} \ln(k_2 A^{\phi/2}), \tag{7.15}$$

where the k's are constants, and S is species richness and $\delta = Q(1 + b) + D + F(a - b) - 9/4$ and $\phi = 1/(1 + F/D)$. As I demonstrated, this expression can be converted so that species richness depends on habitat by assuming $F \le H$ and substituting H for F in equation (7.15).

Note that the Korcak exponents a and b contain the exponent Q, the fractal dimension of resource. To focus attention on habitat area, which should be correlated with total resource density and thus Q, we define resource density to be fixed, such that $Q = \omega H$ such that ω is a fraction that represents the degree to which habitat is occupied by resources. Again, we can assume a 2-dimensional landscape, $D = 2$. Furthermore, for relatively large landscapes of conservation interest, the constants change relatively little with H or Q, and so I focus attention on the exponents in understanding how habitat distribution affects species richness. With these assumptions

$$S = 1 + k_H A^z \ln(k_h A^{1/(2+H)}), \qquad (7.16)$$

where the exponent z contains the influence of resource fractal dimension Q and habitat dimension H on both limiting similarity of species (γ) and on the maximum sampling scale w_{max}. Where γ can be small (high Q relative to H) and w_{max} can be large (high H), the exponent will be large, and species richness at a given landscape extent or area A will be relatively high. The constants k_h and k_H are equivalent to k_1 and k_2 in equation (7.15) but include H rather than F. A closer inspection of z reveals that

$$z = [1/(2(2 + H))][-1/4 + \omega H + H^2(\omega + \omega^2/2 - 3/2)]. \quad (7.17)$$

This very interesting result suggests that species richness depends on habitat area and fragmentation in a complex way. The parameter ω reflects the density of resources *within* habitat, or habitat quality. Analysis of equation (7.17) reveals that the way in which the species-area exponent z responds to an increase in H depends on habitat quality (fig. 7.5). When $\omega < 3/4$, or resources are approximately at a within-habitat density less than $1/\sqrt{x}$, species richness $S = 1$. At $3/4 < \omega < 0.9$, an increase in H corresponds more strongly to an increase in fragmentation than an increase in habitat area, and z declines with increasing H over the range. Thus, for an equivalent area sampled, or equivalent-sized landscape, species richness will decline with increasing habitat dimension H. At high habitat quality $\omega > 0.9$, the exponent z increases with increasing H; and at a given area sampled, species richness will increase with H because increasing H corresponds more strongly to increasing habitat area than to fragmentation. Interestingly, as habitat quality and area increase, the value of z asymptotically approaches

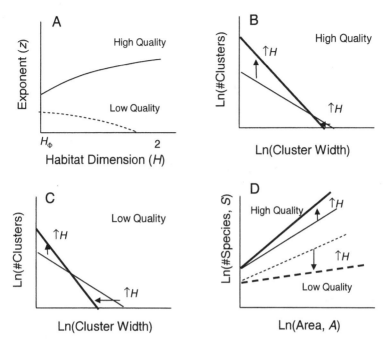

FIGURE 7.5. (A) Hypothetical exponent z of the approximate species-area relationship for large landscapes $S \approx A^z$ over a conservation-relevant range of habitat fractal dimensions $(2 > H > H_\Phi > 1.2)$ for high quality (resource-dense) and low quality (resource-scarce) habitats. (B, C) Shift in the power law frequency distributions of habitat cluster sizes resulting from an increase in H for (B) high quality, and (C) low quality habitats. (D) Resulting species-area curves implied by (A)–(C) for high and low quality habitats.

0.25, the value suggested from the central limit theorem by averaging stochastically varying per capita growth rates of species (May 1975). Under most conditions (as specified in table 5.2), the spatial scaling model predicts the exponent z of species-area relations to range between 0.1 and 0.25, which is within the range typically observed empirically (MacArthur 1972; Storch et al. 2008; Harte et al. 2008).

These qualitative predictions are, to my knowledge, the first explicit linkage between both habitat quality and the scaling exponent of the power law species-area relationship, which applies for the spatial scaling model because the focus is on large landscapes relevant to conservation. The spatial scaling model predicts that, in resource-rich

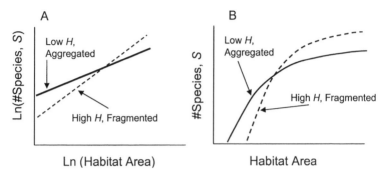

FIGURE 7.6. Hypothetical influence of habitat area and fragmentation on species richness of a guild, plotted in (A) logarithmic space and (B) linear space. Graphs show the greater sensitivity of fragmented habitats to reduction in habitat area but the quicker saturation of species richness with increasing area in aggregated habitats.

environments, fragmentation may increase the number of small habitat clusters that help smaller-scaled species coexist at more similar sizes without a large reduction in the frequency of large habitat clusters needed to support larger-scaled species (fig. 7.5B). In contrast, in resource-poor environments, increasing fragmentation reduces the frequency of large habitat clusters without a sufficiently compensating increase in the number of smaller clusters (fig. 7.5C). Consequently, increasing fragmentation may increase species richness in resource-rich habitats and decrease it in resource-poor habitats, and thus reduce overall species richness (fig. 7.5D). Applied to real-world habitats, species richness may be strongly reduced by fragmentation of unproductive habitats, such as deserts, oligotrophic lakes, polar tundra, etc.; while species richness may be enhanced by moderate fragmentation in more productive habitats like tropical rainforests, eutrophic lakes, and wetlands (Chase and Leibold 2003).

These results lead to further qualitative predictions about how species richness should increase with habitat area. Species richness should increase more rapidly with habitat area or proportion h, in fragmented habitats, with higher H, than in aggregated habitats (fig. 7.6). At low h, what habitat remains is aggregated into sufficiently large clusters to sustain larger-scaled species, and thus, for a given area of habitat, a lower H will sustain more species. In contrast, at high h, smaller-scaled species can coexist at more similar sizes because there is a much higher frequency of smaller habitat clusters but sufficient large clusters to support larger-scaled species. If habitat quality is high enough, species richness may be higher in more fragmented habitats.

These qualitative predictions are based on power law species-area relations. In many cases, studies explore species incidence or species richness as a function of untransformed habitat area rather than the logarithm of habitat area (With and King 1999; Fahrig 2002). If the qualitative predictions in figure 7.6A are plotted vs. untransformed habitat area, one quickly detects "threshold" effects in which species richness or incidence declines much more dramatically below a certain proportion of landscape occupied by habitat (fig. 7.6B). For the exponents ($z = 0.1 - 0.25$) expected for most habitats (fig. 7.5) and observed in the guilds examined in chapter 5 (table 5.1), this sharp downturn will occur in the range of $0.05 < h < 0.15$, or around $h = 10\%$. This is consistent with the results of relatively detailed simulations of species demography, dispersal, and incidence in response to fragmented habitats (Andrén 1994; With and Crist 1995; With and King 1999; Fahrig 2001). This threshold of species loss may occur at a slightly higher h in more fragmented habitats or at a similar threshold of h that is independent of H. This outcome suggests that the trade-off between supporting larger-scaled species in rarer but more aggregated habitats, and more smaller-scaled species in fragmented but more abundant habitats induces robustness in the generic rule of thumb, suggested by landscape fragmentation studies, that perceived catastrophic losses of populations and species should not occur until a habitat has declined to <10% of its original extent. In other words, this 10% rule may hold regardless of whether a habitat is fragmented or not (Fahrig 2002; Radford et al. 2005).

What may matter a great deal for the conservation of species is whether the loss of habitat results in a reduction of habitat quality ω (Vance et al. 2003). One can imagine that higher quality or more productive habitats, such as those on richer soils or with greater access to water, may be preferentially removed, effectively reducing both h and ω. If so, species incidence and richness may decline more precipitously with the loss of habitat than predicted by the 10% rule, and increasing H for a given h may exacerbate the effect. Many fewer studies have documented effects of habitat loss and fragmentation on habitat quality, probably because quality is much more difficult to measure across landscapes and with remote sensing techniques. The analysis in this chapter suggests a renewed urgency for assessing whether habitat loss also induces or is accompanied by a decline in habitat quality, and whether such an effect leads to loss of species at higher habitat proportions than expected.

In contrast with this somewhat gloomy prognosis, the above results of the spatial scaling model suggest that habitat aggregation, or lowering H for a given h, may fairly successfully mitigate species losses under habitat loss (Trzcinski et al. 1999). The relationships in figure 7.6 suggest that preserving a few, large habitat areas may preserve a larger fraction of species than expected for a given proportion of habitat. The qualitative predictions of the spatial scaling model thus reconstitute the old SLOSS (Single Large Or Several Small) debate (Diamond 1975; Simberloff 1982, 1984; Kunin 1997; McCarthy et al. 2006) as one of FLOMS (Few Large or Many Small). In particular, conserving more aggregated habitat may be critical in supporting the larger-scaled, and thus larger-sized species that are often of greatest charismatic interest. In contrast, inherently smaller-scaled species such as arthropods may have thresholds of minimum habitat cluster size that lie far below the scales of resolution at which habitat is mapped for reserve design. Thus the diversity of these species may be related mostly to proportion of habitat rather than habitat fragmentation (Olff and Ritchie 2002; Hoyle and Harborne 2005).

This entire analysis obviously oversimplifies the full range of issues in conservation, as the SLOSS debates and other simplified conservation models have also done in the past. In particular, it assumes that species can readily disperse to habitat clusters, and in this sense fails to capture the "island biogeographic" (MacArthur and Wilson 1967; Hubbell 2001) aspects of habitat fragmentation that arise from limited dispersal across inhospitable habitat that lies between suitable habitat clusters (King and With 2002). Nevertheless, the analysis does point out many ways that resource limitation and scale-dependent constraints and trade-offs might shape the patterns of species richness with habitat loss and fragmentation. The spatial scaling model thus presents several testable alternative hypotheses about how habitat loss and fragmentation might affect the incidence and richness of species, and how different management or reserve design strategies might aid species conservation. These hypotheses are directly connected to the distribution of food and resources on landscapes, and firmly introduce the concept of habitat quality into the discussion.

SPECIES PERSISTENCE AND DIVERSITY ON
REAL LANDSCAPES

It is far beyond the scope of this book to provide a detailed and comprehensive review of studies of the effects of habitat fragmentation on species conservation. However, I offer a few examples in which a fractal geometric approach has been employed as a way of illustrating how the qualitative predictions of the spatial scaling model might be tested. To this end, I explore some examples from my own work (Olff and Ritchie 2002) plus those from some selected field studies of patterns in bird communities (Vickery et al. 1994; Radford et al. 2005).

Han Olff and I analyzed the incidence of particular species and species richness of bird and butterfly (Lepidoptera) communities reported for 36 different landscapes in the Netherlands, each 9 x 9 km (Olff and Ritchie 2002). The habitat of interest is heathland, a type of native shrub/grassland on low-nutrient soils, highly reduced in landscape cover and fragmented by centuries of agriculture and urban development in northern Europe. Heathlands support many species of conservation interest, particularly birds, butterflies, and plants. In each of the 36 landscapes, heathland habitat was digitized from 1:50.000 scale topographic maps that differentiated habitat at 250 m resolution. We calculated fractal dimension of heathland habitat according to the mass fractal method (Milne 1992) with a much coarser resolution than we used in our original paper (fig. 7.1), and found a wide range of values of h and H across the landscapes (fig. 7.3).

To explore the influence of h and H on species incidence and richness, we used data on the occurrence of all higher plant species, breeding birds, and butterflies collected in 1 x 1 km grids nested within these landscapes between 1970 and 1990 in annual surveys by volunteers. The data collection was managed by several Dutch NGOs (SOVON for birds, Vlinderstichting for butterflies, and FLORON for plants). Aggregated data over time was used for this period, as not much has changed to the landscape spatial structure over this time

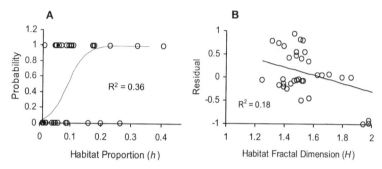

FIGURE 7.7. (A) Logistic regression of the occurrence (0 or 1) of breeding populations of black grouse (*Tetrao tetrix*) with habitat proportion in the landscape (h) for the 36 Dutch landscapes (probability of occurrence = p where the ratio $p/(1 - p) = -3.31 + 35.6h$, $R^2 = 0.36$, $P < 0.004$). (B) Residual variation in occurrence not explained by h is regressed against habitat fractal dimension (H), which for a given h, indicates the degree of habitat fragmentation (*Residual* = $1.43 - 0.869H$, $P = 0.02$) Modified from Olff and Ritchie (2002).

period. We explored patterns for species especially associated with heathlands, since more generalized species that also used heathlands in addition to other habitats would obviously be unlikely to respond only to heathland habitat distribution and cover.

Species Incidence

To test whether species incidence increased with cover and declined with increasing fragmentation, we chose as an example the largest bird species, black grouse (*Tetrao tetrix*). Using logistic regression, we found that the presence or absence of black grouse across the 36 landscapes increased significantly with total cover of heathlands (fig. 7.7A), but decreased with increasing habitat fractal dimension, or fragmentation, given h (fig. 7.7B), as predicted qualitatively by the spatial scaling model (eqn. (7.13)). Further analysis suggests that black grouse occupy landscapes very nearly only with habitat clusters >1 km^2 (Olff and Ritchie 2002), also as predicted by the spatial scaling model (eqn. (7.14)).

These results for the Dutch heathlands are further supported by a pair of field studies. One (Vickery et al. 1994) explored habitat patch

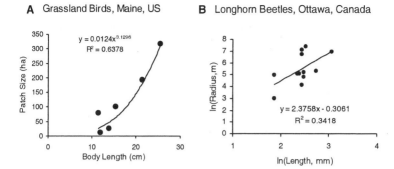

FIGURE 7.8. (A) Minimum grassland patch area in which 6 different grassland specialist bird species of different body lengths were found in a fragmented landscape of coastal grasslands in Maine (Vickery et al. 1994). (B) Scale (radius, m) at which longhorn beetles (Coleoptera: Cerambycidae) exhibit the greatest sensitivity to forest cover in Ontario as a function of body length (Holland et al. 2005).

sizes >1 ha that were used by different specialist grassland bird species associated with coastal dune grasslands spatially isolated by ocean, marshes, and forests in Maine in the northeastern United States. They found that larger bird species occurred only in larger grassland patches. My own analysis of their data shows that the minimum patch area in which each species was found increased with body length, the surrogate for sampling scale I have used in all previous chapters, according to a power law with an exponent of 3.12 (fig. 7.8A), which is very close to the lower value (3.2–4) predicted by the spatial scaling model (eqn. (7.14)).

In another study (Holland et al. 2005), the radius at which forest cover was measured affected the sensitivity of longhorn beetle (Coleoptera: Cerambycidae) incidence to forest cover. Through application of an algorithm that explored the sensitivity of beetle incidence vs. the radius of a circular window at which forest cover is measured, Holland et al. (2005) showed that the radius at which each of 12 generalist species showed their strongest response to forest cover can be interpreted as proportional to the length of the smallest forest patch that will be occupied by beetles of different length. This length scale was significantly correlated with beetle body length, with an exponent of slightly more than 2 (fig. 7.8B). Equation (7.14) relates area sampled by the observer to organism sampling scale, but can be converted to relating observation area length to sampling scale by taking the square root of both sides. This then predicts a scaling exponent of $1 + H/2$, which is likely less than the slope of the ln–ln relationship in figure 7.8B. Although these studies had either small sample sizes (Vickery et al. 1994) or somewhat creative measures of habitat patch scale (cluster size) (Holland et al. 2005), both datasets suggest a power law quantitative relationship between habitat cluster size, incidence, and consumer sampling scale predicted by the spatial scaling model.

Species Richness Patterns

For the Dutch heathlands (Olff and Ritchie 2002), breeding heathland specialist bird species richness increased, as expected, with habitat cover (fig. 7.9A). After controlling for this effect, bird species richness declined significantly with increasing H (fig. 7.9B) This result is consistent with the spatial scaling model's qualitative predictions.

In contrast to the results for birds, butterfly species richness in Dutch heathlands increased with habitat cover, but did not decline significantly with increasing H (fig. 7.9 C, D). The coarse resolution of mapping may mean that all mapped heathland habitat clusters were large enough to support any of the butterfly species. Clusters small enough to not support larger butterflies might not have been mapped or might have been aggregated with other clusters. Alternatively, habitat quality to butterflies may be higher than for birds, such that butterfly richness is insensitive to fragmentation while bird richness declines with fragmentation. Thus the spatial scaling model leads to a fresh alternative hypothesis that might be tested by comparing bird or insect demography across habitat clusters in relation to plant species composition or density, and then using vegetation data to independently assess habitat quality.

Many other studies show similar patterns of diversity with increasing habitat area. Typical studies like these, but with particular relevance to the qualitative predictions of the spatial scaling model, are recent studies of bird species richness in relation to forest cover. In Australia, Radford et al. (2005) found that landscapes with aggregated habitat had higher species richness at low habitat cover (small habitat

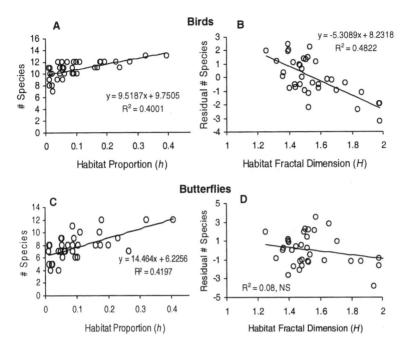

FIGURE 7.9. Regressions of heathland bird species richness (A, B) and butterfly species richness (C, D) with proportion (h) of heathland habitat across 36 Dutch landscapes (9 x 9 km areas, see fig. 7.3). (B, D) Residual variation not explained by h is regressed against fractal dimension (H). Note that residual bird species richness, after accounting for the effects of habitat amount in (A), declines significantly with increasing H, which for a given h indicates the degree of habitat fragmentation. Note also that this pattern for butterflies was not significant ($P > 0.05$). Modified from Olff and Ritchie (2002).

area), while landscapes with fragmented habitat had higher species richness at higher habitat cover (large area) (fig. 7.10). Although the interaction term in the analysis of covariance (with habitat area and aggregation as factors) was not significant, it was nearly so ($P = 0.078$). Furthermore, species richness, when plotted vs. habitat area on a linear scale, declined rapidly at habitat cover <10% for both the aggregated and fragmented landscapes. These two patterns are predicted by the spatial scaling model, and the 10% threshold (With and Crist 1995) has been found in many other studies (Fahrig 2002).

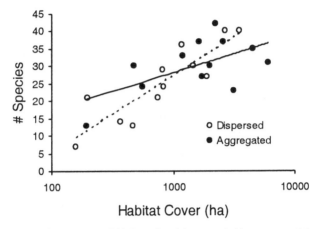

Habitat Cover (ha)

FIGURE 7.10. Response of bird species richness to habitat cover and fragmentation in 100 km² landscapes in southeastern Australia (Radford et al. 2005) that are either aggregated (filled circles) or dispersed (open circles). Both regression lines are significant ($P < 0.05$) and Analysis of Covariance (ANCOVA) shows that the interaction between habitat cover and aggregation is nearly significant ($P = 0.07$).

Similarly, bird species richness increased with habitat area in European landscapes (Storch et al. 2008) with a species-area relationship of slope $z = 0.15$ after dividing the exponent for line transect data, 0.3, by 2 to convert length into area. In addition, the species-area curve over virtually the full range of areas sampled in that study matched very closely with that predicted by a pure fractal distribution of habitat.

These examples and their correspondence to qualitative and quantitative predictions of the spatial scaling model illustrate the possibilities for connecting consumer-resource theory to landscape ecology and biodiversity conservation. As was seen in predictions for the structure and diversity of guilds in local habitats (chapters 5, 6), the critical components of this connection are scales of the observer and organism and heterogeneity, in this case in the form of aggregated or dispersed clusters of habitat. In my view, these examples point to a much richer general, analytical theory of landscape ecology and a plethora of hypotheses to test with further field observations and experimental tests.

SUMMARY

1. The explicit formulation of food and resource distributions developed for the spatial scaling model provides the basis for a model of species incidence and richness in potentially fragmented habitats. The model thus has implications for the conservation of species and biodiversity.
2. Habitats that are fractal can exhibit variation in their fractal dimension H for a given proportional cover of habitat h across a landscape, and increasing H is associated with greater fragmentation for a given h.
3. Criteria from the spatial scaling model derived from its predictions for maximum sampling scale (size) as a function of area and for species-area relationships can be used to derive new predictions for how species incidence and diversity should respond to increasing habitat area, fragmentation, and quality (resource density). Minimum habitat cluster size is predicted to scale directly with species' sampling scale (size) with an exponent of 3.2 in rare habitats to 4 in abundant habitats.
4. Species-area relations depend on both habitat distribution and quality. Aggregation of habitat can mitigate habitat loss, especially in poorer quality or less productive habitats. In contrast, aggregation may lead to lower species richness in more productive habitats.
5. The exponent z, for power law species-area relationships depends as strongly on habitat quality as area, but may be relatively insensitive to habitat fragmentation except in the poorest and richest quality habitats. The exponent is predicted to lie between 0.1 and 0.15 for most habitats, considerably lower than the exponent derived from a lognormal distribution of abundances (May 1975).
6. Empirical studies of species incidence and richness across landscapes with different proportions and aggregation of habitat generally support the qualitative predictions of the spatial scaling model, and suggest that it may provide considerable insights and hypotheses about the effects of landscape management on species extinction and persistence and biodiversity.

CHAPTER EIGHT

Testing the Model

Now that the reader is armed (and potentially dangerous!) with the spatial scaling model, he or she might ask, how do I test it? I have tested the major assumptions and predictions of the model at virtually every opportunity throughout this book, and while these datasets are certainly not exhaustive, the limited number of studies that simultaneously measure size, resource consumption, abundance and species richness strongly constrain tests at this time. Nevertheless, these tests represent just the beginning of what needs to be done to validate the model for situations where its assumptions apply, and to compare its qualitative predictions with those of other models (e.g., Huston 1994; Tilman 1994; Hubbell 2001; Volkov et al. 2003; Harte et al. 2008).

First of all, the model yields many qualitative predictions about the nature of consumer diets, food or habitat patch use, size distribution shapes (chapter 5), and diversity patterns (chapter 6). These qualitative predictions represent the fruit of a general analytical theory and serve as the basis for countless ways the model can be tested without fitting parameters.

Unfortunately, many models of community structure and diversity yield similar qualitative predictions, such as when the lognormal distribution and the spatial scaling model yield very similar species-abundance distributions (chapter 6). In this chapter, I discuss the issues and possible protocols in designing quantitative tests of the spatial scaling model against data and against those of other theories of community structure. In all the previous chapters I have demonstrated specific examples of these quantitative tests, along with certain adaptive considerations that make them possible; so this chapter will deal more with the philosophical assumptions an empirical ecologist might need to make in quantitatively testing the model.

A major issue in testing the model is that there are a relatively large number of parameters in the somewhat complicated expressions for size ratio, γ, maximum sampling scale w_{max}, and other key outputs of the model. However, closer inspection reveals only six parameters: extent x, fractal dimension of the environment D, fractal dimension of food F, fractal dimension of resources Q, number of sampling volumes in the renewal time for the resource v, and B_1, the scale-independent coefficient of the relationship between resource requirements and scale. For community patterns in relation to habitat distributions, one can substitute a habitat dimension H for F (see chapter 7). In an ideal world, one would like to be able to measure each of the parameters in the spatial scaling model independently and make a fully deductive prediction for each pattern. This is straightforward for 4 parameters. In selecting the community for study, the observer typically chooses the geometry, or dimension, of the environment D, scale of resolution ε, and thus an extent or scale of observation x. In addition, it is straightforward in most cases to measure the fractal dimension of food F, resources Q, or habitat H, following established methods (Johnson et al. 1996).

However, there are two parameters, v and B_1, that are very difficult to measure a priori. The parameter v is difficult because the renewal rate, and thus renewal time, of the resource is difficult to measure. Furthermore, so many different environmental factors control metabolic rate and thus resource requirements, that B_1 is also difficult to estimate a priori. These difficulties mean that there are typically two "free" parameters in the spatial scaling model that cannot easily be specified in advance and may require some sort of fitting.

Having two free parameters does not mean that the model cannot be tested and compared with other models. In this regard, the spatial scaling model, if D, x, F, and Q are specified, is no different in degrees of freedom from linear regression and has few disadvantages when compared with other community models, such as neutral theory (Volkov et al. 2003). The spatial scaling model is quite testable because of the large number of simultaneous qualitative predictions it makes. For example, in this book I "fit" the model to best explain patterns of size ratios of species adjacently ranked in size, and then (without any additional parameters or fitting), took these fitted parameters to predict, major patterns in species abundance for seven different guilds. This method sacrifices the independence of a test of one pattern in order to estimate certain difficult parameters and then independently test many others.

Another issue is, given the large number of parameter inputs, can the model be rejected as an explanation for a set of data? I think I have demonstrated quite clearly in chapters 5 and 6 that it can. The model can completely fail to fit the size ratio patterns for some guilds (pelagic ciliate herbivores, the tintinnids), in which case the hypothesis that the community is structured by size-dependent resource partitioning is summarily rejected. Furthermore, the model may predict size structure and a species abundance distribution for a local community, such as vascular plants in Minnesota, quite well (chapter 5), but be rejected as the model predicting species-area relationships (chapter 6).

It is important to explore in more detail how to estimate the four a priori parameters. First, extent x is almost always set by the observer's choice of scale and resolution (the unit of measuring space). For example for Cedar Creek grasshoppers, the resolution was set at 1 mm and the extent was thus the length of one side of a plot, $x = 9000$ mm (see table 5.1). For dung beetles, resolution was also 1 mm, but x was the length of a dung patch (table 5.1) from which beetles extracted different-sized dung balls (chapters 3, 4). My view is that the resolution ε should correspond to the unit of measurement at which the length of different species is measured.

The second parameter that is easily estimated is D. Over large extents in terrestrial and aquatic benthic environments, such as the conservation landscapes discussed in chapter 7, this will almost always be 2. This was appropriate for the Serengeti large mammal community (table 5.1). In aquatic pelagic environments or other spaces with significant vertical components like tropical rainforests, D may be much greater than 2 and can approach 3 for small species like pelagic unicells or even forest birds if the horizontal extent x is reasonably close to the vertical height y. The vertical dimension Y should be estimated following the arguments outlined in chapter 5 (see eqn. 5.20) and illustrated in figure 8.1. Again, the vertical extent y can be expressed as a fraction δ of x, where $\delta = y/x$, and the vertical dimension $Y = 1 + \ln(\delta)/\ln(x)$. Adding Y to the width and length dimensions (each equal to 1) yields $D = 2 + Y$. For pelagic species, such as protist ciliate tintinnids (Dolan et al. 2007; table 5.1), Y may very well approach 1, and thus D = 3.

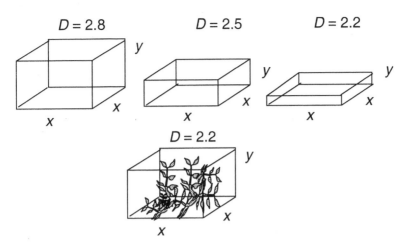

FIGURE 8.1. Hypothetical environmental geometries that determine the environmental dimension D in the spatial scaling model. Along the top row, the vertical height y decreases relative to the isometric horizontal extent x, which reduces D according to the equation $D = 1 + \ln(\delta)/\ln(x)$, where $\delta = y/x$. The lower figure illustrates D for organisms restricted to plant surfaces in a space with the same height y as the upper left illustration.

Structure in the environment poses complications. As Morse et al. (1985) and more recently McAbendroth et al. (2005) point out, the actual volume, and thus geometrical dimension, of the environment is reduced if species are restricted to surfaces in space, such as insects on leaves, or microbes on the surfaces of soil particles (Finlay and Fenchel 2001). Consequently, arthropods on leaf surfaces of a forest in which forest height y might be as large as x, and thus $Y = 1$ might actually experience only $D \leq 2.25$, because considerable empty space exists between leaf or twig surfaces. Another example would be the difference in geometry experienced by bark-feeding, raptorial, and aerial insectivorous birds living in the same rainforest (table 5.1). Bark-feeding birds are restricted to hunting along plant and ground surfaces, which in 3-dimensional space have $D \leq 2.2$ (Morse et al. 1985). In contrast, aerial insectivores search the entire air column from the forest floor to the canopy, an environment very close to 3-dimensional. Any test of the spatial scaling model needs to carefully consider whether consumer organisms are restricted to surfaces and to measure the potential fractal properties of these surfaces to estimate D.

The next two parameters of interest are the two fractal dimensions, which can also be estimated in most systems, if somewhat less confidently. Terrestrial food is often readily measurable as plant leaves, fruit, insects, dung, litter, etc. Even aquatic organisms, such as fish, algae, krill, etc., can be "observed" with hydroacoustic equipment (e.g., Tsuda 1995; Seuront and Lagadeuc 2001; Warren et al. 2005) or fluorescence techniques (Waters and Mitchell 2002; Waters et al. 2003). Microbes can also be observed with fluorescence techniques. If such measurements can be made in a spatially explicit way, such as a map or line intercept (Voss 1986, see chapters 2 and 7), then F can be measured directly (Hastings and Sugihara 1993). Even in the absence of spatially explicit data, such as might apply to microbial environments, F for an assumed fractal distribution can be estimated from the mean food density m, where $F = D + \ln(m)/\ln(x)$ (see chapters 2, 4, and 5). Likewise Q, which is often very tedious to measure in a spatially explicit manner (Olff and Ritchie 2002), can be estimated from the mean resource concentration r of food as $Q = F + \ln(r)/\ln(x)$. For a further discussion of appropriate use of different methods to estimate fractal dimensions, see Hastings and Sugihara (1993), Johnson et al. (1996), and Halley et al. (2004).

Once x, D, F, and Q are estimated, the two parameters v and B_1 remain, and as I mentioned before, these are very difficult to measure. However, having two fitted parameters is not particularly alarming since this would make the spatial scaling model no different than a linear regression. Again, the multiple qualitative predictions of the spatial scaling model make it possible that one set of observations can be fitted in order to predict other patterns with no further fitted parameters. These can be fitted with least squares (table 5.2) or maximum likelihood methods (Bolker 2008).

MODEL SENSITIVITY

Another major issue in testing the model is the sensitivity of model predictions to errors in estimation or fit. A sensitivity analysis of the spatial scaling model for predicting species richness (fig. 8.2) reveals, not surprisingly, that the model is most sensitive to changes in the exponents that define the dimensions and geometry of the environment,

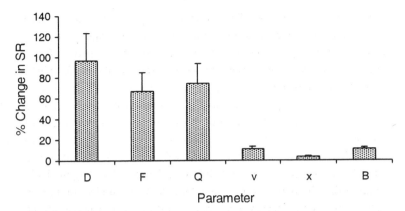

FIGURE 8.2. Mean (±SE) percent change in species richness (SR) predicted by the spatial scaling model from 100 simulated 20% changes in each parameter of the model. The original D, F, Q, v, x, and B_1 in each simulation were selected from uniform distributions of parameters within reasonable ranges (table 5.1).

D, F, and Q. Predictions are hypersensitive to these parameters in that a 3- to 5-fold greater response results from a given proportional change in these exponents. Interestingly, species richness was most sensitive to increasing D, which may reflect an increasing vertical dimension (see chapter 5). The good news is that these parameters can be estimated a priori with reasonable precision (<10% error). In contrast, predictions change by about only half of the magnitude of change in the two fitted parameters v and B_1. This relative lack of sensitivity in the two fitted parameters suggests that algorithms for fitting these two parameters are likely to be stable, and confidence intervals for fits are likely to be relatively tight. This latter point is important in potentially comparing the quantitative predictions of this model with those of others.

DATA TYPES, SAMPLING, AND EXPERIMENTS

In this book, I used data from my own published and unpublished research, as well as data from other published sources. Each data source has certain strengths and weaknesses in testing the spatial scaling or other community models. The literature is replete with data on size structure of sets of various taxa, and offers the opportunity for meta-analyses of multiple communities from multiple environments. However, I found that these datasets are poorly resolved as to the food habits or trophic position of different species, and as I will show fairly convincingly in chapter 9, lumping related taxa with dissimilar food habits can lead to major distortions in size structure and unfair tests of the model. Furthermore, it may be difficult to assess the fractal dimensions of the environment, food, and/or resources with literature data, since these were not collected with testing the spatial scaling model in mind. A final limitation is that the volume or area sampled for a given data set is often unclear, which leaves questions as to whether all the species reported in a publication are from a single geometric space or represent the sum of several samples. Sampling from multiple sites can have similar effects on the observed distributions as sampling from across species with very different food habits.

A better approach is to design a field study with the spatial scaling model in mind. Ideally, one would want to have a clear idea of the geometric space to be sampled, such as a specific research plot or defined sampling area. The fractal dimension of the environment, food, and resources could presumably be measured with standard methods. This would leave, as discussed above, only 2 parameters to be fit to data, and allow the least ambiguity in fractal dimensions, to which the model is most highly sensitive. In addition, a nested set of spatial extents could be sampled, which would allow a species-area relationship to be determined. This approach was used successfully for the grasshopper and plant communities from Cedar Creek Natural History Area in Minnesota.

By far the best way to test the spatial scaling model is with experiments. For a wide variety of experimental systems from microbial microcosms to spatial distributions of nutrients at 100 m extents, the fractal dimension of food and/or resources could be manipulated in replicated treatments. Replicates could be "seeded" with a large range of species, which presumably would compete according to the mechanisms implied in chapter 5 and sort down to a set of species whose body sizes, abundances, and species richness could be compared with those predicted by the model for a known food and/or resource fractal dimension. Alternatively, in field-scale experiments with say, arthropods, plants, or small vertebrates such as fish, one might let species from the "regional" pool colonize the different fractal dimension treatments. This approach can provide an especially powerful test when the abundance or relative abundance of species in the regional pool is known, because then it would be possible to generate predictions from neutral community theory (see chapter 5 for examples and chapter 9 for further discussion) to compare with observed data.

As an example of an experimental test with dung beetles in South Africa, I subdivided a large supply of fresh white rhino dung into smaller piles of various sizes (10–100 cm in diameter) and then placed these equidistant from three different large white rhino middens, which served as a source of dung beetle species. I resampled inside and under these "experimental" dung piles and then compared the minimum and maximum size and number of species found in each (chapter 4). Although I lacked sufficient time in the field to do such an experiment, it would have been easily possible to collect fresh rhino dung and distribute it in different spatial arrays that would correspond to different food fractal dimensions. Using dung from different mammalian herbivores (see fig. 4.11) could change the nutrient concentration and so that could have been varied as well. Replicating these different configurations would have allowed a direct experimental test of how changing food fractal dimension should change community structure.

CONCLUSIONS

The spatial scaling model can be explicitly and quantitatively tested in a variety of ways from literature, survey, and experimental data, provided some information is available beyond the abundances and sizes of a set of related species. The key to a successful test is to identify a set of trophically similar species, regardless of taxa, and to carefully measure or define the geometry of the environment and the food and resources within it. Communities from multiple samples in space and time should not be averaged. The most powerful tests will come from experimental manipulation of the geometry of the environment, food, and/or resources and be accompanied by quantitative tests of the spatial scaling model's predictions. Knowledge of species' relative abundances in a larger regional environment would allow a direct comparison of expected community structure from neutral theory with that of the spatial scaling model and the observed data. Only after such experimental tests of the theory, can ecologists have sufficient confidence that agreement between data and models is driven by the mechanisms presumed in the models.

SUMMARY

1. The spatial scaling model can be tested in a variety of ways with data from the literature, designed surveys, or best of all, experiments that manipulate the fractal dimension parameters of the model.
2. Fitting the model to data potentially requires tuning six parameters, which is a high number; but typically, one can define or measure four of these a priori. Such a course makes the model no more parameter-rich than a 2-factor linear regression.
3. Many of the model's predictions can be tested without any adjustable parameters by fitting the model to one pattern, such as the body size distribution, and then directly computing all the other patterns, such as productivity-diversity, species abundance distributions, species-area curves, and/or species diversity vs. habitat loss.
4. The model's predictions are, as expected, most sensitive to changes in the geometric dimensions of the environment, food, and resources, and relatively insensitive to changes in extent, movement rates, and the renormalization coefficient for resource requirements. Fortunately, these dimensions can often be measured a priori and thus not be treated as adjustable parameters.
5. The most powerful tests will come from experimental manipulation of the geometry of the environment, food, and/or resources and be accompanied by quantitative tests of the spatial scaling model's predictions. Knowledge of species' relative abundances in a larger regional environment would allow a direct comparison of expected community structure from neutral theory with that of the spatial scaling model and the observed data.

CHAPTER NINE

Perspectives, Caveats, and Conclusions

The past four chapters present a series of models that, when combined, form the "spatial scaling model" of community structure. This model explores how heterogeneity in the distribution and packaging (as food) of a resource might influence individuals, populations, and communities of consumers whose fitness is limited by that resource. The assumption of heterogeneous resources generates the basis for selective foraging, resource partitioning, and competitive coexistence of community structure: as I mention repeatedly, this model collapses to the classic consumer-resource models of Tilman (1976, 1982) and Chase and Leibold (2003) if resources are distributed randomly rather than heterogeneously. However, there are many ways to describe heterogeneity. Why does the approach of fractal geometry, optimized resource acquisition, and competitive coexistence lead to such a comprehensive theory that seems to predict some community attributes so well? How is it that the assumptions used to make the model create such a synthetic connection between individual behavior and biodiversity, that can predict, unlike almost every other niche-based community model, virtually all the major patterns of biodiversity (Ritchie and Olff 1999)?

Five major assumed attributes of the consumer-resource system drive the spatial scaling model's structure and qualitative predictions.

1. Resources and their food packaging are distributed as fractals, such that the frequency distribution of different cluster sizes of food and resource can be described with fractal geometry.

2. Resources are limiting, which means that all species in a community are limited by the same resource and thus represent "guilds."

3. Consumers maximize their resource intake under all conditions—this assumption leads directly to selective foraging for particular sets of food or resource clusters, and provides an evolutionary argument for resource partitioning that depends on the sampling scale of the consumers.

4. There is no dispersal limitation—all consumers can encounter exclusively used clusters of the appropriate food cluster size and resource concentration.

5. Competition is the principal species interaction determining coexistence.

When these assumptions are reasonably met, the model predicts observed resource selection, exclusive resources, limiting similarity in size, maximum and minimum size, species richness, and even species abundance strikingly well. In many cases, such as for minimum food cluster size and resource concentration used by herbivores and dung beetles (chapter 4), the predictions match observed data better than previous models. For many community attributes, the model's qualitative predictions are completely novel in that they represent the first success at *explicitly* connecting optimized resource use with population size and community structure. Because these predictions are novel, at this point I cannot compare them with those from alternative niche-based models.

In this chapter, I conclude by highlighting major insights from the successes of the model in predicting resource partitioning and community structure of guilds. I follow this by revisiting the major assumptions of the spatial scaling model and evaluating these contributions to the model qualitative predictions. I discuss more deeply my justification of these assumptions, and reveal further insights from evaluating the model in light of observed data. Finally I end the book by pointing to the future, where the dynamics of coexistence outlined in this book become a piece of a much larger theory that combines niche-assembly (Chase and Leibold 2003), exploitation of temporal and spatial heterogeneity and the storage effect (Chesson 2000; Snyder and Chesson 2004), and neutral dispersal-assembly processes (Hubbell 2001; Etienne and Olff 2004; Volkov et al. 2003; Harte et al. 2005) into something approaching a "unified" theory of biodiversity and biogeography.

INSIGHTS FROM THE SPATIAL SCALING MODEL

The success in connecting foraging to community structure through the fractal geometry of heterogeneity gives some powerful new insights into how resources may structure communities. These insights come from both the logical connections between resource-maximizing foraging, population dynamics, species coexistence, and community structure and from the strong success in predicting observed resource use, size structure, and major diversity patterns in a select group of guilds for which there is sufficient information. As one would expect for any simple model, some observed data deviate from model quantitative predictions, and these deviations also provide valuable "signals" of processes and mechanisms other than resource limitation and the partitioning of heterogeneity that contribute to community structure. In this section, I explore some of these major insights.

The spatial scaling model strongly re-emphasizes body size, which is perhaps the best surrogate (as a measurable key axis of niche differentiation) for consumer sampling scale. The importance of size was originally envisioned by G. Evelyn Hutchinson (1959) and Robert MacArthur (1958, 1965) (MacArthur and Pianka 1966), and re-iterated many times since (e.g., Schoener 1971, 1976; Abrams 1980a; Grant 1986; Belovsky 1997; Enquist 2001; Brown et al. 2002). However, the model fully formalizes the concept of size as a key trait for coexistence with an explicit, general mechanism to show how body size determines resource use, how food cluster size and resource concentration trade-off with size, and how this contributes to species coexistence. It is in fact the fractal nature of heterogeneity that makes size so important, as size represents the consumer's scale of measurement, and consumers with different sampling scales encounter different densities of food and resources (chapters 2 and 3). These differences in encounter lead to different resource-maximizing sets of food clusters of different size and resource concentration for consumers of different size; and in turn, these different sets of food clusters determine shared and exclusive resources and species coexistence (chapters 4 and 5). If food and/

or resources are randomly or uniformly distributed, as is assumed by the classic consumer-resource and Lotka-Volterra models, then size confers no difference in access or use of resources and cannot, in this context, act as an axis of differentiation for species coexistence. Thus the degree to which heterogeneity is fractal may influence the importance of size in structuring communities.

The issue of coexistence brings up a second major insight: competition for heterogeneous resources may occur through the dynamics of competition for shared and exclusive resources (Schoener 1976, 1978) and not through the differential rates of consumption of different resource types (Ritchie 2002), as is described by Lotka-Volterra competition or classic consumer-resource models (MacArthur 1969; Tilman 1982). As I discussed in depth in chapter 5, the concept of competition for both shared and exclusive resources is not new—it was developed in the 1970s (Schoener 1976, 1978). With the exception of work by Gary Belovsky (1986, 1997) and Jon Chase (1996), it was largely ignored as a mechanism of competition. I believe this happened because so many ecologists were focused on competition coefficients as descriptors of the impact of one species on another, and they assumed that non-overlapping use of resources resulted in a competitive interaction where each species has a constant per capita impact on the other. Schoener (1976) showed quite clearly that this was not the case; species that use different sets of resources effectively have both exclusive and shared resources (fig. 9.1), and decidedly variable per capita effects on other species (Ritchie 2002). Competition coefficients and Lotka-Volterra dynamics would describe interactions for shared but not exclusive resources. Given the number of non-overlapping resource spectra measured during the proverbial "halcyon days" of competition in the 1970s and early 1980s (Schoener 1983; Connell 1983), competition for shared and exclusive resources may be much more common than ecologists may be willing to admit.

A third major insight is that partitioning of spatial heterogeneity may be, in many cases, the most important mechanism explaining coexistence and relative abundance of species in guilds. For the guilds I used as examples in this book, the signal of size structure and declining abundance in direct proportion to size-dependent resource requirements was very strong (chapter 5) in 5 of the 7 cases. Classical consumer-resource models do not predict this signal because size is not considered

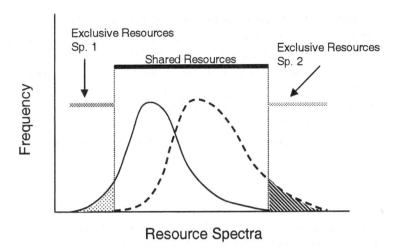

FIGURE 9.1. Hypothetical resource utilization spectra, analogous to food cluster size or cluster resource concentration, for 2 competing species (Sp. 1, thin curve, Sp. 2, dashed curve). Regions of the resource spectra that do not overlap (lightly shaded for Sp. 1, hatched for Sp. 2) confer exclusive resources, while regions of overlap are shared resources. Lotka-Volterra competition would only occur for shared resources, and if sufficiently abundant, exclusive resources might allow for coexistence.

important. Models of dispersal limitation and community drift (Hubbell 2001; Volkov et al. 2003) are statistical and have no required signal of size or resource requirements, although such additions are possible (Etienne and Olff 2004). Other mechanisms, such as the diversity-modifying effects of predators or herbivores may depend on size (Sinclair et al. 2003; Brose et al. 2005), but the way in which this size dependence would structure a diverse community has not yet been worked out. For now, the partitioning of fractal resources described in the spatial scaling model seems to provide a strong "first approximation" of the size structure and abundance distribution of guilds of species.

AN EVALUATION OF KEY ASSUMPTIONS

The remarkable success of the spatial scaling model in connecting consumer foraging to abundance and community structure for certain guilds of species comes with the burden of the five major assumptions previously listed. Almost certainly some communities, such as Rohde's (2001) fish parasites and Dolan's (2000, 2005) tintinnid protists, do not meet these assumptions and exhibit different size structure, abundance, and diversity patterns than predicted by the spatial scaling model. Below, I consider each of the five assumptions in detail as a way of exploring the points of departure from resource partitioning of fractal resources by a guild of species, and thus the utility of the spatial scaling model as a hypothesis for community and diversity patterns.

Fractal Geometry

A major question is: How much do the results depend on the assumption of fractal geometry (chapter 2)? Fractal geometry is the necessary mathematical framework to describe a "neutral" assumption about heterogeneity; its pattern is explicitly the same at all scales of observation. Fractal geometry, however, invokes a decidedly "non-neutral" assumption about the scale at which organisms sample the environment, since the amount of a fractal substance measured in the environment depends on the scale at which it is measured (Mandelbrot 1982). This "sampling" or "measurement" scale-dependence, applied to consumer organisms, drives most of the qualitative predictions of the resulting set of foraging (chapter 3), population growth (chapters 3–5), and species interaction (chapters 5,6) models that comprise the spatial scaling model. As such, the model is as much a prediction of variation in sampling scale among consumers as it is a model of consumer-resource population dynamics and coexistence. For this reason, the model's qualitative predictions quite strongly depend on the assumption of fractal geometry, as it is this explicit mathematical framework that drives qualitative predictions of maximum and minimum sampling scale and minimum scale ratios that define niche space and species coexistence. Given the results of comparing observed communities to model quantitative predictions, fractal geometry seems to be a useful first order description of the spatial distribution of resources (Ostling et al. 2003; Green et al. 2003; Halley et al. 2004).

The table can be turned to ask: What if the distribution is decidedly not fractal? In the case of randomly or uniformly distributed resources, the answer is easy: the model collapses into the classic $R*$

competition model for a single limiting resource, in which case one species with the lowest R^* will exclude all other species. The more complex case is one in which there is one or more "breaks" in the pattern of heterogeneity as the scale of observation increases (Schooley and Wiens 2001; Allen and Holling 2002). Such breaks would impose a hierarchical structure to space and create likely designations for "local" dynamics, which depend on the food and resource distributions at small extents. "Regional" dynamics would emerge from the interaction of locally coexisting species among different localities, which are linked according to the spatial pattern that applies at a larger scale of observation. Mathematically, this can be dealt with by making the exponents (fractal dimensions) of the scaling laws imbedded in the theory become functions of scale. This greatly complicates the mathematics and eliminates most of the possibilities for analytical solutions, but keeps the essential element of scale in the model and in the predictions. Such a model could be iteratively solved to obtain quantitative predictions for particular size structure, abundance, or diversity patterns of interest.

A different approach is to embrace the scale break in heterogeneity patterns across scales and separate dynamics into "local" and "regional" environments. This approach would essentially construct "competitive metacommunities" (Amarasekare and Nisbet 2001; Amarasekare et al. 2004; Klausmeier and Tilman 2002; Mouquet and Loreau 2003) where coexistence of many species is possible at regional scales because conditions vary across localities, but only a few species can coexist locally because of the constraints of classical consumer-resource interactions. The consequence of scale breaks for community structure and biodiversity is an exciting current topic of research that has been little explored.

Guilds

The spatial scaling model strictly assumes that all species are limited by the same resource in the same type of packaging, such as herbivores, carnivores, autotrophs, etc. This restricts any qualitative predictions to guilds, a concept that dates to Darwin (1859) and Huxley (1942) in the context of the evolution of body size and character displacement. Guilds can sometimes be difficult to define because not all species consume all the same food or even operate entirely on the same trophic level (Hawkins and MacMahon 1989), but the definition I used (chapter 4) is any set of species that are likely limited by the same food and resources. For animals this seems to provide good results for herbivores, detritivores, and carnivores. Other guilds also seem easy to define, such as specialized vertebrate parasites (Rohde 2001) and plants (chapters 5 and 6).

Larger problems in defining guilds arise when species undergo significant ontogenetic life history and diet shifts (Werner and Gilliam 1984; Werner 1994). I experienced this in defining guilds of grasshoppers at Cedar Creek, Minnesota. The correct approach from a practical, functional point of view is to treat each ontogenetic stage as if it were a separate species. This means that a significant fraction of the "species" in a community might be different-sized stages of the same actual species. Alternatively, the competitive dynamics of the community might change over time as different species attain different sizes. For example, in the Cedar Creek grasshopper community, several species' sizes are the mean sizes of nymphs at the time of sampling, not the eventual adult size. With some flexibility and common sense, it seems possible to accommodate ontogenetic niche and diet shifts in groups of organisms where they predominate, such as fish, amphibians, aquatic insects, etc.

From the perspective of testing the model, a more serious problem is that many community samples are organized on the basis of taxa rather than resource-limitation. For example, the literature is full of papers that try to interpret mammal, bird, or arthropod community patterns without trying to distinguish different guilds within those taxa. One can readily see that lumping guilds is likely to give different patterns than for individual guilds. For example, five different guilds of Amazonian forest birds (Terborgh et al. 1990) each exhibit diversity-size distributions that are strongly left-skewed (fig. 9.2A). The size class corresponding to maximum species richness differs considerably among guilds, so that when all species are combined as "forest birds," a lognormal diversity–size distribution results (fig. 9.2B). Without recognizing this, one might be tempted to argue, on the basis of taxa-specific rather than guild-specific patterns, that the data reject the hypothesis of fractal resource partitioning as an explanation for community structure.

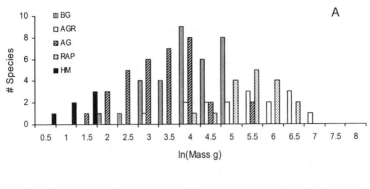

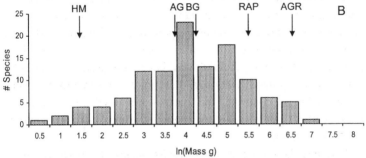

FIGURE 9.2. (A) Frequency distribution of species by different body size classes (ln(mass) in g) for different guilds of Amazonian forest birds (Terborgh et al. 1990); bark gleaners and woodpeckers (BG), arboreal granivores (AGR), arboreal gleaners (AG), raptors (RAP), and hummingbirds (HM). (B) Frequency distribution of species among body sizes for all 5 guilds combined, with arrows indicating size class of peak diversity for each guild.

The pattern observed in figure 9.2 is likely to be repeated for any large taxa composed of two or more guilds that differ in their maximum size, as was the case for these birds. Lumping different guilds can also produce strongly right-skewed distributions, observed for large-scale distributions of mammals (Brown et al. 1993) and birds (Schoener 1984; Maurer and Brown 1988), when most guilds, such as insectivores, granivores, frugivores, etc. have small maximum sizes and a few have large maximum sizes (carnivores, folivores). Clearly the patterns we expect for taxonomically defined communities may be very different than patterns expected for guilds, especially if a taxon contains a diversity of guilds.

Optimal Foraging

Unlike existing consumer-resource models, the fractal foraging model developed in chapters 3 and 4 assumes that consumers are selective. This assumption is certainly supported by a massive literature. More controversial is the model's assumption that consumers maximize intake of a single resource, the premise behind the classical work on optimal foraging from the mid 1960s to the mid 1980s (Stephens and Krebs 1986). Despite arguments that organisms do not have enough information or that foraging decisions reflect a balancing of alternative currencies (Ward 1992), repeated experiments have shown that organisms often select sets of food items, prey, or particles that are strongly correlated with sets that would maximize a single currency (Stephens and Krebs 1986; Belovsky 1986). Although a majority of studies focused on energy maximization, and energy would certainly be a valid "resource" in the context of the spatial scaling model, many consumers may select patches, clusters, or diets that maximize other currencies, such as protein.

Certainly these often-repeated results suggest that resource use by most organisms is much more consistent with optimizing intake of some currency rather than random foraging. In incorporating the presumed evolutionary pressure of maximizing resource intake, the spatial scaling model predicts differences among species in the sets of food and resources used, where random foraging, as assumed by classical consumer-resource models, does not. Regardless of the details of currencies, information, and cognition of consumers, an optimal foraging assumption provides a much better approximation of nature than does random foraging. Consequently, the spatial scaling model provides a clear, testable hypothesis of how such evolutionarily shaped foraging might lead to the structuring of communities. As with the issue of size discussed previously, the connection of optimal foraging to community structure fulfills the original vision and purpose for optimal foraging proposed by Emlen (1966) and MacArthur and Pianka (1966).

The assumption of fractal geometry in fact allows an analytical solution to the optimal foraging behavior of consumers in the spatial scaling model. The frequency distribution of different food and resource cluster sizes is a power law under fractal geometry, and such power laws are readily integrated mathematically (chapter 3). Thus the imposition of a specific geometry sufficiently constrains the potential frequency distribution of food clusters with different "profitability" to allow analytical qualitative predictions of limiting similarity and species coexistence. An optimal foraging assumption also introduces an inherent flexibility in the behavior of consumers to match changes in the supply of food and resources (chapter 4), yet species ranks in minimum food cluster size P^* and resource concentration R^* remain unchanged despite big changes in absolute differences in these thresholds. This flexibility makes the model robust enough to predict, closely in many cases, resource partitioning and size structure for the wide variety of guild types I explored in chapters 5 and 6. All these benefits to an optimal foraging approach in my mind greatly outweigh the risks of predicting behavior that may not be quite attainable by consumers.

Dispersal Limitation

The spatial scaling model assumes no dispersal limitation by any of the consumer species in the community. Consequently, all species are assumed to have probabilities approaching 1 of detecting food clusters of acceptable size and resource concentration during the renewal time of resources in the environment. This assumption may be highly appropriate for mobile animals observed at relatively small scales. However, many organisms are clearly limited in their ability to arrive at particular points in space and to exploit particular environmental conditions. Plants in particular have been repeatedly demonstrated to be dispersal limited (Tilman 1994; Karst et al. 2005), and many communities such as fish parasites (Rohde 2001), and marine plankton (Dolan 2000, 2005; Dolan et al. 2007) all exhibit the high species turnover across space and low sample frequency that are consistent with dispersal limitation.

Fundamentally, dispersal limitation implies that any given species, at a given density and mobility v, may not be able to exploit all available clusters of food and resources during resource renewal time τ. Such failure means that there are resources left available for another ecologically equivalent species to exploit. If so, then a community limited by the same resources might have at least some species that coexist through a dispersal-limitation dynamic rather than by resource partitioning. In certain cases, the mobility of the consumer relative to the dispersion of resources may be so weak that many equivalent species can be maintained through the process of community drift (Hubbell 2001, 2006).

The possibility of dispersal limitation suggests that any given community may sit somewhere on a continuum between no dispersal limitation and complete dependence of coexistence on resource partitioning vs. intense dispersal limitation and almost no dependence on resource partitioning (Tilman 1994). In chapter 5, the size structure of guilds such as Serengeti herbivorous mammals and Cedar Creek grasshoppers seem to be very well described by size-dependent resource partitioning, whereas that of marine herbivorous protists (tintinnid ciliates) corresponds much more closely to a random pattern. In the case of protists, the species abundance and beta diversity patterns are highly consistent with the qualitative predictions of neutral theory and the mechanism of dispersal-limited ecological drift (Dolan et al. 2007). Other guilds, such as dung beetles and plants, show size structure consistent with resource partitioning but abundance and diversity patterns that include the possible signal of dispersal limitation. These examples may illustrate the continuum of possibilities. The spatial scaling model therefore provides, for the first time, an explicit niche assembly-based prediction of size structure, species abundance, and species richness to complement that of neutral or dispersal-limited assembly (Karst et al. 2005; Alonso et al. 2006).

The spatial scaling model provides some additional insight into when communities might begin to be influenced by dispersal limitation. In the derivation of consumer-resource intake in chapter 4, equation (4.2), there is the parameter p_F, the probability that a food cluster will be encountered in v sampling volumes of size w^D:

$$p_F = 1 - \left(1 - x^{F-D}\right)^{w^D}, \tag{9.1}$$

which thus far I have assumed would be very close to 1. Simple inspection shows that as food (and the resources in it) becomes scarce (F approaches 0), as consumer sampling scale w decreases or as the scale of observation x increases, p_F will decline (fig. 9.3). Although not strictly the same as the colonization or dispersal rate in a population

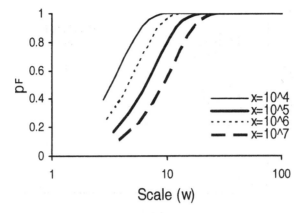

FIGURE 9.3. Response curves of the probability of finding a food cluster for consumers of different scale (w) at different landscape extents (x). Similar responses occur with decreasing F, reflecting lower density or increasing food aggregation, or with increasing D.

dynamic model, this probability reflects the tendency for individual consumers to fail to "successfully disperse" to food clusters that they would otherwise exploit. This encounter probability suggests some very important things. First, species are more likely to be dispersal limited over larger scales of observation. Second, smaller-sized species are more likely to be dispersal limited. Third, dispersal limitation is more likely as food becomes more rare or highly aggregated in space (F decreases), especially as the environmental dimension D increases from 2 to 3.

These testable qualitative predictions suggest that ecological equivalents might occur more frequently among small size classes, with more rare, aggregated resources, in more 3-dimensional environments, and/or if communities are sampled at a very large scale. Such dispersal-limited population dynamics and coexistence is of course not directly described or accounted for in the spatial scaling model. A simple qualitative prediction of its influence on community structure comes from assuming simplistically, that each species within a specified size class (operationally defined by $\gamma(w)$, the limiting similarity) can, in time τ, occupy p_F of possible food clusters, such that the number of equivalent species in that size class is proportional to $1/p_F$. If so, we might find a greater number of species than expected in small body size classes.

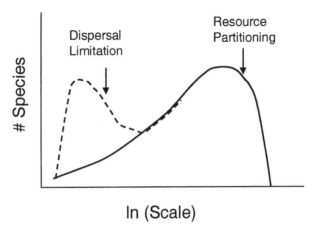

FIGURE 9.4. Theoretical consequences of dispersal limitation on the diversity-scale relationship induced by decreasing the probability of food encounter p_F.

The size ratio-size relationships for the spatial scaling model from chapter 5 imply a left-skewed distribution of species richness for different body size (or sampling scale) classes (figs. 5.6–5.8). Because they must be farther apart in size on a logarithmic scale to coexist, fewer species will occur in smaller size classes. Examples of such left-skewed distributions are those of the Amazonian bird guilds in figure 9.2. If ecological equivalents are much more likely to occur because of dispersal limitation in smaller body size classes, then we might observe many more species in the smallest size classes than expected from the spatial scaling model (fig. 9.4). Thus, the signal of dispersal limitation, even under strong resource competition within a guild, might show a bi-modal diversity-size distribution, with peaks of diversity in the smallest and largest size classes. Such bi-modal distributions would be more evident as p_F declines, so we would expect them under conditions of rare, aggregated resources, larger spatial extents, and/or a significant vertical dimension.

As examples of the patterns one might expect, I made two comparisons in figure 9.5. First, I compared two bird guilds from the Peruvian Amazon (Terborgh et al. 1990): one forages on relatively rare and highly aggregated fruit resources distributed within trees, a habitat with an extensive vertical dimension (arboreal frugivores); and the other forages

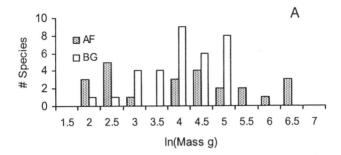

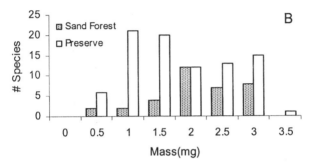

FIGURE 9.5. Diversity-size distributions for (A) Amazonian birds of two different guilds, arboreal frugivores (AF) and bark gleaners and woodpeckers (BG) (Terborgh et al. 1990), and (B) dung beetles in Mkuzi Preserve, South Africa (Doube 1991) from a specific habitat (Sand Forest) and the entire preserve.

on the same trees but on a much more dispersed and potentially more abundant resource (insectivorous bark gleaners and woodpeckers). The two guilds have significantly different diversity-size distributions (Kolmogorov-Smirnov test), and the arboreal frugivore guild has a significantly bi-modal distribution (DIP test, Giacomelli et al. 1971) similar to what we might expect with dispersal limitation of the smallest size classes. The reason for the difference in diversity-size patterns between the two guilds is completely speculative at this point, since we have no direct measures of F for fruit or bark insects. Nevertheless, the distributions are clearly different in a way that is at least consistent with the hypothesis that small members of the arboreal frugivore guild may experience probabilities of encounter with fruit (p_F) << 1, thus allowing many more species of very similar smaller body size to coexist.

The second example compares the diversity-size distribution of dung beetles from a different South African game preserve, Mkuzi Game Reserve, KwaZulu Natal, sampled with dung traps spread over the entire reserve (340 km^2 vs. from a single specialized habitat (sand forest, 16 km^2). At the smaller scale, single habitat sample, beetles exhibited a strongly left-skewed distribution (fig 9.5B). At the larger scale, and as would be predicted by dispersal limitation of smaller size classes coupled with resource competition and partitioning, beetles exhibited a significantly bimodal distribution of diversity. Again, it is speculative to suppose that increasing scale by a factor of 20 is sufficient to induce dispersal limitation, but the results are consistent with what might be expected from the spatial scaling model.

This "thought experiment" with influences on p_F and possible dispersal limitation proves relatively little. However, it does suggest that resource partitioning and dispersal limitation (or other mechanisms that can produce ecological drift) can occur at the same time and in the same community. The competition–colonization trade-off dynamic (Pacala and Tilman 1994; Tilman 1994; Klausmeier and Tilman 2002) is an example of a model where both processes operate simultaneously, although it assumes no spatial heterogeneity and has not been used to predict community structure patterns. Nevertheless, the spatial scaling model should provide a useful reference point against which to compare neutral theory and other models that include dispersal limitation. From such comparisons, dispersal assembly and niche assembly mechanisms can possibly be reconciled and synthesized (Leibold and McPeek 2006).

Other Interactions Determining Community Structure

The spatial scaling model describes the competitive interactions of species and does not explicitly include predation or speciation, two other mechanisms that have been shown to influence diversity. Many theoretical (e.g., Holt et al. 1994; Chase and Leibold 2003), classic, and recent empirical studies (Paine 1966; reviews by Sih et al. 1985; Olff and Ritchie 1998; Schmitz and Suttle 2001; Chase 2003) show that predation can structure abundance and diversity of communities. Likewise, speciation rates in some groups, such as cichlid fishes in African Rift Valley lakes (Albertson et al. 1999) and tropical forest birds and butterflies (Cardillo 1999), are fast enough to generate considerable diversity of species that have similar sizes, food habits, and habitats. Other mechanisms, such as disturbance (Connell 1978; Huston 1994) and sympatric speciation through sexual selection (Gavrilets 2004) might also determine species richness. So it seems unlikely, perhaps, that a model of interspecific competition should accurately predict resource use, abundance, and community structure.

Predation could be implicitly incorporated into the spatial scaling model through the loss rate term B, originally presented in chapter 3. As Schoener (1976) showed, the population dynamic version of resource requirements includes both the resources lost to maintenance and those lost to death, which could easily include predation. From the perspective of partitioning fractal resources, predation would be most important and cause the greatest change in predictions if it caused the loss rate to scale differently. For example, heavier predation on smaller size classes (Sinclair et al. 2003; Brose et al. 2004, 2005) could lead the scaling relationship of loss rate with size to have an exponent β lower than 9/4 (chapter 5), because the added burden of predation mortality of smaller-scaled species would increase the loss rate at smaller scales. If so, the scaling exponent for loss rate and the limiting similarity in scale would both decrease (inspect eqn. (5.17)), allowing more species to coexist under the same supply rate of food and resources. Under such a scenario, predators would act as "keystone" species and increase diversity. In contrast, selective predation on the largest size classes (Sogard 1997; Mehner and Thiel 1999) could steepen the scaling relationship between loss rate and size and reduce diversity. Yet another scenario would be that the effects of predation are distributed evenly across size classes and therefore predation would not affect the exponent β but would just increase the constant B_1. This would result again in an increase in limiting similarity and a decline in species richness for the same food and resource supply conditions.

The magnitude and scaling of predator impacts on loss rate, and the general influence of predators on dynamics and competitive coexistence of prey of different size is virtually unstudied (Chase et al. 2003). The spatial scaling model, however, clearly can accommodate different scenarios and selective feeding patterns of predators that are prevalent in the data (Chase et al. 2003) to predict realistic responses of guilds to predation. I did not develop this further in this book simply because there are virtually no empirical studies of how predation affects the scaling of loss rate, and so a detailed analysis of model qualitative predictions would likely leave the reader still uninformed about how predators would modify the size-dependent coexistence of species for shared and exclusive resources.

Speciation is the other major alternative mechanism of maintaining diversity (Otte and Endler 1989), and much exciting research is underway to discern what controls it (Cardillo 1999; Gavrilets 2004; Allen et al. 2006). Speciation is the major source of diversity in neutral models (Hubbell 2001; Volkov et al. 2003) since there is no mechanism to cause species to disappear other than, by chance, low abundance. However, the dynamics modeled in this book are relatively fast compared to the rate of speciation, so that speciation may shape the pool of available species that can exploit available resources, but is unlikely to determine the structure of the community that might coexist under current resource conditions. Put another way, speciation occurs too slowly to generate different-scaled (or sized) individuals to exploit resources available in the short-term. However, if dispersal limitation increases in sufficiently fragmented habitats or over large enough environmental extents, as I have hypothesized, then speciation may ultimately determine the number of possible ecological equivalents of similar scale or body size that could coexist. Greater speciation rates might increase the number of ecological equivalents. Equivalents might enhance the diversity in certain environments, like tropical rainforests (Cardillo 1999; Hubbell 2001), the Mediterranean fynbos ecosystem in South Africa (Linder and Hardy 2004), or cichlid fishes in the African Rift Valley (Albertson et al. 1999). These have all maintained endemic species that emerged from speciation events over millions of years. The relative importance of speciation, dispersal limitation, and local competitive interactions that can scale up to a region (chapter 6) have simply not been tested or studied, and this is likely a very fruitful area of future research.

PROSPECTS FOR THE FUTURE

Aside from testing the spatial scaling model's assumptions and qualitative predictions, which has been discussed in detail in the sections above, the research climate is ripe for three major future efforts.

1. The model can serve as a "niche assembly" template against which to compare observed patterns of body size structure, abundance, and diversity of guilds of species against alternative theoretical models, particularly models that assume dispersal limitation.

2. The model can be explored much more thoroughly than I have done here to generate and test qualitative predictions about major patterns in size structure, abundance, species richness, and the linkages between these.

3. The model serves as a "niche assembly" entry point into a more unified theory of community structure and diversity, that will combine other major and successful (in certain contexts) theories such as neutral theory (Alonso et al. 2006), information theory (Harte et al. 2005, 2008; Shipley et al. 2007), competition–colonization trade-off (Pacala and Tilman 1994; Tilman 1994) and other metacommunity approaches (Leibold et al. 2004), and metabolic theory (Brown et al. 2004).

In chapter 6 and in this chapter I discussed in detail how the spatial scaling model provides, in many cases, distinct qualitative predictions from theories based on different mechanisms of community organization, so the first point just re-emphasizes this.

The second point recognizes that the predictions and data explored throughout the book are but the beginning of, in my mind, a rejuvenated study of mechanistic community ecology. Much more needs to be learned about the major assumptions and qualitative predictions of the spatial scaling model. For example, a "Top Ten List" of questions might be:

1. How good is fractal geometry as an approximation of real material distributions?

2. How does sampling scale w vary with traits other than body size?

3. How does the relationship between size and selective foraging change under different food and resource supplies?

4. Does the niche space predicted by the spatial scaling model represent a "capacity" for species that might correlate generally with species richness (Currie et al. 2004)?

5. How often, and with what outcome, do species compete for shared and exclusive resources, and is the partitioning of these resources driven by variation in body size?

6. Is there evidence of "energetic equivalence" or roughly equal resource availability per unit area or volume for species of different size?

7. How do observed patterns of similarity in size change with food abundance, resource concentration, habitat dimensions, mobility, and factors that change metabolic rate and thus resource requirements?

8. How do dispersal limitation and resource competition combine to influence community structure and species-area relationships?

9. How does resource renewal time vs. supplied mass as components of productivity vary across systems and does it influence diversity in the manner predicted?

10. How does habitat destruction, which presumably would reduce the environmental dimension D as well as food and resource dimensions (or increase their clustering), affect species richness and species abundance patterns?

The spatial scaling model suggests these and many more interesting questions. The model rejuvenates the study of "deterministic" community ecology because, for the first time, we have an explicit way to connect consumer-resource and coexistence mechanisms to actual patterns of community structure. As discussed in the previous section, deviations from the model's assumptions are also likely to lead to important insights, new models, and a deeper understanding of how resource limitation alone can lead to a rich set of community patterns rather than competitive exclusion.

In the context of other new work on the mechanisms that control diversity, the spatial scaling model suggests a potential vision of how different theories and their component models can be linked. Such unification would help dissolve the current, and likely continuing, controversies about niche vs. dispersal assembly and trait-based vs. trait-independent species coexistence that contrast between models. I offer one possible vision here in figure 9.6, based on the results of a 2007 workshop at the Santa Fe Institute focused on unifying biodiversity

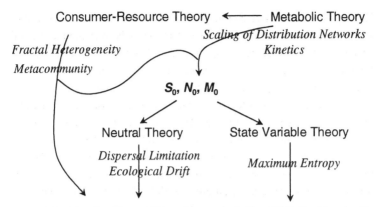

Figure 9.6. Hypothetical framework for unifying major theories of ecological community structure. Arrows show how major theories inform other theories, bold text represents key predictions, and italics highlight key models that generate predictions. Consumer-resource and metabolic theory together can predict key large-scale (at area A_0) community parameters (total species S_0, total abundance N_0, and total mass M_0).

theories. In my view, there are currently four major active theories of community ecology, each of which have specific component models that predict. Two are trait-independent theories that assume that differences in traits among species do not matter in determining their relative abundance or spatial distribution: neutral theory and state variable theory. The other two are theories in which species traits, particularly size and resource demand, directly determine species membership, abundance, and diversity: consumer-resource theory and metabolic theory. Each of these four theories feature specific models, such as the new spatial scaling model in this book and metacommunity models within consumer-resource theory, or the information theory/maximum entropy state variable approach (Shipley et al. 2007; Harte et al. 2008; Dewar and Porte 2008).

Most ecologists view these theories as separate and virtually mutually exclusive entities. However, it seems possible that these theories, through their key component models, might be unified because they are connected by key parameters. Neutral (Hubbell 2001; Volkov et al. 2003; Etienne and Olff 2004) and spatial macroecology (Kunin 1998;

Hartley et al. 2005; Harte et al. 2005, 2008) theories require as inputs some known or initial total number of individuals (N_0) and number of species (S_0) at a specified large area (A_0). With these inputs, the two theories can predict the relative abundance and number of species found at some smaller area, often with quite astounding success. In the case of spatial macroecology, this can also be extended to predict endemics (species found in only a subset of smaller areas). The spatial scaling model and emerging metacommunity models (Leibold et al. 2004), I believe, represent an evolution of consumer-resource theory, the dominant conceptual paradigm in niche-based community ecology for the past three decades (Leibold and McPeek 2006). This "evolved" consumer-resource theory predicts how environmental influences such as space and resource supply, coupled with the constraints on the demand for resources, such as temperature and body size as predicted by metabolic theory (Brown et al. 2004), to determine N_0 and S_0 at a specified A_0. These predictions can be used to constrain the trait-independent processes of neutral or spatial macroecology models. Alternatively, the spatially informed consumer-resource models can make independent quantitative predictions of community structure and abundance that can be compared with the trait-independent models and with observed data.

This unified framework of theories and their component models (fig. 9.6) allows for a wealth of community-structuring processes to be considered simultaneously. Yet different hypotheses, each associated with the predominance of a single process or combination of processes, can be generated and tested with observed data. The spatial scaling model represents a first step in generating consumer-resource models that are sensitive to spatial heterogeneity in resource availability. Such models now offer the possibility of understanding community structure as the net result of niche assembly and dispersal assembly or from a combination of trait-dependent vs. trait-independent processes. For guilds of mobile species with reasonably abundant resources, such as detritivores like dung beetles, generalist mammalian or insect herbivores, competition and resource partitioning for shared and exclusive resources may be the dominant determinant of community structure. For other guilds, such as tropical forest trees and planktonic herbivores, dispersal assembly mechanisms may predominate. For many other groups, the multiplicity of mechanisms implied by the unified framework of theories may ultimately determine species abundance and diversity. This is a future with many exciting hypotheses to test and hopefully little in the way of polemic polarization around particular ideas. Surely the success of the model in this book and others developed recently provides an optimistic view of understanding that most difficult of subjects, community ecology (Lawton 1999). It is, indeed, not time to "move on" but to move forward.

SUMMARY

1. The spatial scaling model of community structure successfully predicts a wide array of resource partitioning, size structure, and the abundance and diversity of species for guilds of mobile consumers such as herbivores, mammals, and grasshoppers (Orthoptera), decomposers such as dung beetles, or carnivores such as birds.

2. The success of the model derives from its assumptions: that limiting resources are packaged in other material and distributed as fractals, that consumers maximize their resource intake subject to constraints, that there is no dispersal limitation, and that competition is the primary species interaction determining coexistence.

3. These assumptions are not always met, and many interesting departures from the qualitative predictions of the model result if they are not. In particular, the interaction between size-dependent resource partitioning and dispersal limitation, or between partitioning and predation, offer interesting possibilities for future research.

4. Aggregating the patterns for different guilds within a habitat, or across different habitats for the same guild, yield different species abundance and diversity patterns than for a single guild in one habitat. Familiar patterns, such as lognormal (Preston 1962) or right-skewed (Brown et al. 1993) species abundance or diversity distributions, for communities organized by taxonomy or sampled across multiple habitats may emerge by accumulating guilds with different food and resource densities, and thus different maximum and minimum sizes.

5. The spatial scaling model may serve as a potential hypothesis of niche assembly of communities that can be compared with observed data and quantitative predictions of other theories. To this end, the

model has six parameters, but four of these can be measured prior to tests. Despite its complexity, the model therefore has only two relatively insensitive, adjustable parameters to fit to data, and is easier to test than one would expect.

6. The spatial scaling model represents an improvement in consumer-resource theory that permits a diverse array of qualitative predictions about community structure and diversity. The model, when combined with predictions from metabolic theory (Brown et al. 2004), can predict the inputs to trait-independent theories, such as neutral (Alonso et al. 2006) and state variable/maximum entropy (Harte et al. 2008) theories, or can offer its own alternative predictions of community structure and diversity.

7. The statistical tools developed over the past two decades for quantifying heterogeneity, which have facilitated the spatial scaling and other new models of community structure and diversity, offer optimism for the development of a deductive and predictive science of community ecology.

Appendix

Summary of Model Parameters

Descriptions of parameters used in the development of the spatial scaling model of community structure, with the chapter in which they are defined or introduced included in parentheses.

LENGTHS AND TIME

ε Scale of resolution for the system, typically 1mm or smaller (chapter 2)

x Extent or length of the window (or landscape) through which the observer views the system, measured in number of units of length ε (chapter 2)

r Fraction of extent x of unit length ε in a Cantor carpet (chapter 2)

w Length, in number of units of resolution, of the scale at which the consumer samples the environment (chapter 2)

v Number of sampling units of length w sampled by a consumer in time τ (chapter 3)

τ Time for a unit of food or resource to be re-supplied after it is consumed (chapter 3)

L Length of an organism in units of ε (chapter 3)

VOLUMES AND DENSITIES

N Number of units occupied by material of interest in a Cantor carpet (chapter 2) or population density of a consumer species (chapters 3–7).

V Total volume sampled by a consumer in time τ (chapter 3)

q Density, or proportion of space occupied by resources in a landscape (or observation window) of extent x (chapter 2)

m Density, or proportion of space occupied by food in a landscape (or observation window) of extent x (chapter 4)

A Area observed in length units of ε (chapter 2)

$C(.)$ Volume of resources consumed in time τ by a consumer, as a function of different variables (chapter 3)

B Resource consumption rate required for maintenance and replacement reproduction (chapter 3)

G Cluster size of resources (chapters 2, 3)

P Cluster size of food (chapter 4)

R Resource concentration of food cluster (chapter 4)

I, I_{ij} Supply of shared resources across landscape of extent x in time τ (chapters 3, 5)

E_j Supply of exclusive resources across landscape of extent x in time τ (chapter 5)

M Mass of organism or material (chapter 2)

DIMENSIONS

D Dimension of landscape or volume observed, usually Euclidean (1, 2, or 3), but can accommodate a partial vertical dimension (chapters 2, 5)

Q Fractal dimension of resources (chapter 2)

F Fractal dimension of food or resource packaging (chapter 4)

H Fractal dimension of habitat relevant to conservation questions (chapter 7)

Y Partial vertical dimension of D, that accounts for the vertical extent of a habitat within an area of much larger horizontal extent (chapter 5)

CONSTANTS, COEFFICIENTS, AND DIMENSIONLESS NUMBERS

$\Delta \square$ Ratio of sizes at which clusters are arbitrarily differentiated by the observer (chapter 2)

c_G Constant reflecting the increase in probability of encountering a *resource* cluster with an arbitrary incremental increase in cluster size, as given by Δ (chapter 2)

c_P — Constant reflecting the increase in probability of encountering a *food* cluster with an arbitrary incremental increase in cluster size, as given by Δ (chapter 2)

c — Product of c_G and c_P (chapter 4)

p_E — Probability of encountering a unit of *resource* of any size in a sampling volume of length w in chapter 3, and probability of encountering a cluster of exclusive resource in chapter 5

p_F — Probability of encountering a unit of *food* of any size in a sampling volume of length w (chapters 4, 9)

p_O — Probability that a given sampling volume is empty (chapters 2, 3)

θ — Coefficient (dimensionless) for the scaling of minimum *food cluster size* with w (chapter 4)

σ — Coefficient (dimensionless) for the scaling of minimum *resource concentration* cluster size with w (chapter 4)

B_1 — Coefficient or renormalization constant for the scaling of resource requirements (B) with w (chapter 5)

β_{ij} — Ratio of resource consumption for species j relative to i (chapter 5)

$\gamma_{ij}(w_j)$ — Ratio of limiting similarity in scale between species j and species i, where $w_i > w_j$, as a function of the scale of species j (chapter 5)

Φ — Habitat fragmentation index, or number of habitat clusters or patches/habitat area (chapter 7)

References

Abrams, P.A. 1975. Limiting similarity and the form of the competition coefficient. *Theoretical Population Biology* 8:356–375.

Abrams, P.A. 1980a. Resource partitioning and interspecific competition in a tropical hermit crab community. *Oecologia* 46:365–379.

Abrams, P.A. 1980b. Consumer functional response and competition in consumer-resource systems. *Theoretical Population Biology* 17:80–102.

Abrams, P.A. 1988. Resource productivity-consumer species diversity: simple models of competition in spatially heterogeneous environments. *Ecology* 69:1418–1433.

Abrams, P.A. 1996. Limits to the similarity of competitors under hierarchical lottery competition. *American Naturalist* 148:211–219.

Abrams, P.A. 1999. Is predator-mediated coexistence possible in unstable systems? *Ecology* 80: 608–621.

Adams, D. 2007. Organization of Plethodon salamander communities: guild-based community assembly. *Ecology* 88:1292–1300.

Adkison, G.P., and Gleeson, S.K. 2004. Forest understory vegetation along a productivity gradient. *Journal of the Torrey Botanical Society* 131:32–44.

Albertson, R.C., Markert, J.A., Danley, P.D., Kocher, T.D. 1999. Phylogeny of a rapidly evolving clade: the cichlid fishes of Lake Malawi, East Africa. *Proceedings of the National Academy of Sciences* 96:5107–5110.

Allen, A.P., Brown, J.H., and Gillooly, J. F. 2002. Global biodiversity, biochemical kinetics and the energetic-equivalence rule. *Science* 297:1545–1548.

Allen, A.P., Gillooly, J. F., Savage, V., and Brown, J.H. 2006. Kinetic effects of temperature on rates of genetic divergence and speciation. *Proceedings of the National Academy of Sciences* 103:9130–9135.

Allen, C.R., and Holling, C.S. 2002. Cross-scale structure and scale breaks in ecosystems and other complex systems. *Ecosystems* 5:315–318.

Allen, T.F.H., and Hoekstra, T.W. 1992. *Toward a unified ecology*. Columbia University Press, New York.

Allen, T.F.H., and Starr, T.B. 1982. *Hierarchy: perspectives for ecological complexity*. University of Chicago Press, Chicago.

Alonso, D., Etienne, R.S., McKane, R.J. 2006. The merits of neutral theory. *Trends in Ecology and Evolution* 21:451–457.

Amarasekare P., Hoopes M.F., Mouquet N. and Holyoak M. 2004. Mechanisms of coexistence in competitive metacommunities. *The American Naturalist* 164:310–326.

Amarasekare, P., and Nisbet, R.M. 2001. Spatial heterogeneity, source-sink dynamics and the local coexistence of competing species. *The American Naturalist* 158:572–584.

Anderson, K.J., Jetz, W.W. 2005. The broad scale ecology of energy expenditure of endotherms. *Ecology Letters* 8:310–318.

Anderson, T.M., Ritchie, M.E., Mayemba, E., Eby, S., Grace, J.B., and McNaughton, S.J. 2007. Forage nutritive quality in the Serengeti ecosystem: the roles of fire and herbivory. *American Naturalist* 170:343–357.

Andrén, H. 1994. Effects of habitat fragmentation on birds and mammals in landscapes with different proportions of suitable habitat: a review. *Oikos* 71:355–366.

Armstrong R.A., and McGehee R. 1980. Competitive exclusion. *American Naturalist* 115:151–170.

Armstrong R.A., McGehee R. 1976. Coexistence of species competing for shared resources. *Theoretical Population Biology* 9:317–328.

Barnsley, M.F. 1988. *Fractals everywhere*. Academic Press, San Diego, California.

Bartoli, F., Phillipy, R., Doirisse, M., Niquet S., and Dubuit, M. 1991. Structure and self-similarity in silty and sandy soils: the fractal approach. *Journal of Soil Science* 42:167–185.

Battley, E.H. 1987. *Energetics of microbial growth*. John Wiley, New York.

Bell, G. 2001. Neutral macroecology. *Science* 293:2413–2418.

Bell, R.H.V. 1970. The use of the herb layer by grazing ungulates in the Serengeti. Pages 111–123 in Watson, A., ed. *Animal populations in relation to their food resources*. Blackwell, Oxford, UK.

Belovsky, G.E. 1984. Moose and snowshoe hare competition and a mechanistic explanation from foraging theory. *Oecologia* 61:150–159.

Belovsky, G.E. 1986. Generalist herbivore foraging and its role in competitive interactions. *American Zoologist* 25:51–69.

Belovsky, G.E. 1997. Optimal foraging and community structure: the allometry of herbivore food selection and competition. *Evolutionary Ecology* 11:641–672.

Belovsky, G.E., and Slade J.B. 1995. Dynamics of two Montana grasshopper populations: relationships among weather, food abundance and intraspecific competition. *Oecologia* 101:383–396.

Berendse, F., Elberse, W.T., and Geerts, R.H.M.E. 1992. Competition and nitrogen loss from plants in grassland ecosystems. *Ecology* 73: 46–53.

Blackburn, T.M. and Gaston, K.J. 1999. The relationship between animal abundance and body size: A review of the mechanisms. *Advances in Ecological Research* 28:181–210.

Bolker, B. 2008. *Maximum likelihood methods in the R computing environment*. Princeton University Press, Princeton, New Jersey.

Brose, U., Ostling A., Harrison, K., and Martinez, N.D. 2004. Unified spatial scaling of species and their trophic interactions. *Nature* 428:167–171.

Brose, U., Berlow, E.L., and Martinez, N.D. 2005. Scaling up keystone effects from simple to complex ecological networks. *Ecology Letters* 8:1317–1325.

Brown, J.H. 1975. Geographical ecology of desert rodents. Pages 315–341 in Cody, M.L., Diamond, J.M., eds. *Ecology and Evolution of Communities*. Harvard University Press, Cambridge, Massachusetts.

Brown A.M., and Rose, C.H. 1969. Effects of temperature on composition and cell volume of *Candida utilis*. *Journal of Bacteriology* 97:261–272.

Brown J.H. 1981. Two decades of homage to Santa Rosalia: toward a general theory of diversity. *American Zoologist* 21:877–888.

Brown J.H., Gillooly J.F., Allen A.P., Savage V.M. and West G. 2004. Toward a metabolic theory of ecology. *Ecology* 85:1771–1789.

Brown, J.H., Kelt, D.A., and Fox, B. J. 2002. Assembly rules and competition in desert rodents. *American Naturalist* 160:815–818.

Brown, J.H. 1995. *Macroecology*. University of Chicago Press, Chicago.

Brown, J.H., Fox B.J., and Kelt, D.A. 2000. Assemby rules: desert rodent communities are structured at scales from local to continental. *American Naturalist* 156:314–321.

Brown, J.H., Marquet, P.A. and Taper, M.L. 1993. The evolution of body size-consequences of an energetic definition of fitness. *American Naturalist* 142:573–584.

Burnham, K.P., and Anderson D.R. 2002. *Model selection and multi-model inference: a practical information-theoretic approach*. Springer, New York.

Calder, W.A. 1984. *Size, function, and life history*. Harvard University Press, Cambridge Massachusetts.

Capizzi, D., and Luiselli, L. 1996. Feeding relationships and competitive interactions between phylogenetically unrelated predators (owls and snakes). *Acta Oecologica* 17: 265–284.

Cardillo, M. 1999. Latitude and rates of diversification in birds and butterflies. *Proceedings of the Royal Society B*. 266:1221–1225.

Charnov E.L. 1976. Optimal foraging, the marginal value theorem. *Theoretical Population Biology* 9:129–136.

Charnov, E.L. 1994. *Life history invariants*. Oxford University, Oxford UK.

Chase, J.M. 1996. Differential competitive interactions and the included niche: An experimental analysis with grasshoppers. *Oikos* 76:103–112.

Chase, J.M. 1999. Food web effects of prey size refugia: Variable interactions and alternative stable equilibria. *American Naturalist* 154:559–570.

Chase, J.M. 2003. Strong and weak trophic cascades along a productivity gradient. *Oikos* 101:187–195.

Chase, J.M., Abrams, P.A., Grover. J.P., Diehl, S., Chesson, P., Holt, R.D., Richards, S.A., and Nisbet, R.M. 2003. The interaction between predation and competition: a review and synthesis. *Ecology Letters* 5:302–315.

Chase, J.M., and Belovsky, G.E. 1994. Experimental evidence for the included niche. *American Naturalist* 143:514–527.

Chase, J.M. and Leibold, M.A. 2002. Spatial scale dictates the productivity-diversity relationship. *Nature* 416:427–430.

Chase, J.M., and Leibold, M.A. 2003. *Ecological niches*. University of Chicago Press, Chicago.

Chase, J.M., and Ryberg, W.A. 2004. Connectivity, scale-dependence, and the productivity-diversity relationship. *Ecology Letters* 7:676–683.

Chase, J.M., Wilson, W.G., and Richards, S.A. 2001. Foraging trade-offs and resource patchiness: theory and experiments with a freshwater snail community. *Ecology Letters* 4:304–312.

Chesson, P. 1994. Multispecies competition in variable environments. *Theoretical Population Biology* 45:227–276.

Chesson, P. 2000. General theory of competitive coexistence in spatially-varying environments. *Theoretical Population Biology* 58:211–237.

Cody, M.L. and Diamond, J.L., eds. 1975. *Ecology and evolution of communities*. Harvard University, Cambridge, Massachusetts.

Connell, J.H. 1978. Diversity in tropical rain forests and coral reefs. *Science* 199:1302–1310.

Connell, J.H. 1980. Diversity and the coevolution of competitors, or the ghost of competition past. Oikos 35:131–138.

Connell, J.H. 1983. On the prevalence and relative importance of interspecific competition: Evidence from field experiments. *American Naturalist* 122:661–696.

Cooper, S.D., Barmuta, L., Sarnelle, O., Kratz, K., and Diehl, S. 1997. Quantifying spatial heterogeneity in streams. *Journal of the North American Benthological Society* 16:174–188.

Cornell, H.V. 1999. Unsaturation and regional influences on species richness in ecological communities: A review of the evidence. *Ecoscience* 6:303–315.

Cornell, H.V., and Lawton, J.H. 1992. Species interactions, local and regional processes, and limits to the richness of ecological communities: A theoretical perspective. *Journal of Animal Ecology,* 61(1):1–12.

Covich, A. 1972. Ecological economics of seed consumption by *Peromyscus*: a graphical model of resource substitution. *Transactions of the Connecticut Academy of Arts and Sciences* 44:71–93.

Crawley, M. 1997. *Plant ecology*. 2nd edition. Blackwell, Oxford, UK.

Cuddington, K., and Yodzis, P. 2002. Predator-prey dynamics and movement in fractal environments. *American Naturalist* 160:119–134.

Currie, D.J., Mittelbach, G.G., Cornell, H.V., Field, R., Guegan, J.F., Hawkins, B.A., Kaufman, D.M., Kerr, J.T., Oberdorff, T., O'Brien, E., and Turner, J.R.G. 2004. Predictions and tests of climate-based hypotheses of broad-scale variation in taxonomic richness. *Ecology Letters* 7:1121–1134.

Damuth, J.D. 1981. Population density and body size in mammals. *Nature* 290: 699–700.

Damuth, J.D. 1991. On size and abundance. *Nature* 351:268–269.

Damuth, J.D. 1998. A common rule for animals and plants. *Nature* 395: 115–116.

Damuth, J.D. 2007. A macroevolutionary explanation for energy equivalence in the scaling of body size and population density. *American Naturalist* 169:621–632.

Darwin, C. 1859. *On the origin of species*. Oxford, UK.

Davidson, D.W., and Samson D.A. 1985. Granivory in the Chihuahuan desert: interactions within and between trophic levels. *Ecology* 66: 486–502.

DeAngelis, D.L. 1992. *The dynamics of nutrient cycling and food webs*. Chapman and Hall, New York.

Dewar, R.C., Porte, A. 2008. Statistical mechanics unifies different ecological patterns. *Journal of Theoretical Biology* in press.

Diamond, J.M. 1978. Niche shifts and the rediscovery of interspecific competition. *American Scientist* 66:322–331.

Diamond, J.M., and Case, T.J., eds. 1986. *Community ecology: principles and applications*. Harper and Row, New York.

Dodson, S.I., Arnott, S.E., and Cottingham, K.L. 2000. The relationship in lake communities between primary productivity and species richness. *Ecology* 81:2662–2679.

Dolan, J.R. 2000. Tintinnid ciliate diversity in the Mediterranean Sea: longitudinal patterns related to water-column structure in late spring–early summer. *Aquatic Microbial Ecology* 22:68–70.

Dolan, J.R. 2005. An introduction to the biogeography of aquatic microbes. *Aquatic Microbial Ecology* 41:39–48.

Dolan, J.R., Ritchie, M.E. and Ras, J. 2007. The 'neutral' community structure of planktonic herbivores, tintinnid ciliates of the microzooplankton, across the SE Tropical Pacific Ocean. *Biogeosciences* 4:297–310.

Dolan, J.R., Ritchie, M.E., Tunin, A., and Pizay, M. 2008. Dynamics of core and occasional species in the marine plankton: tintinnid ciliates in the N.W. Mediterranean Sea. *Journal of Biogeography* 36:887–895.

Doube, B.M. 1991. Dung beetles of southern Africa. Pages 133–155 in Hanski, I. and Cambefort, Y., eds. *Dung beetle ecology*. Princeton University, Princeton, New Jersey.

Dunbar, R.I.M. 1990. Environmental determinants of fecundity in klipspringer (*Oreotragus oreotragus*). *African Journal of Ecology* 28:307–313.

Elser, J.J., Sterner, R.W., Gorokhova, E., Fagan, W.F., Markow, T.A., Cotner, J.B., Harrison, J.F., Hobbie, S.E., Odell, G.M., and Weider, L.W. 2000. Biological stoichiometry from genes to ecosystems. *Ecology Letters* 3:540–550.

Elton, C.S. 1958. *The ecology of invasions by plants and animals*. University of Chicago Press, Chicago.

Emlen, J.M. 1966. The role of time and energy in food preference. *American Naturalist* 100:611–617.

Enquist, B.J. 2001. Invariant scaling relations across tree-dominant communities. *Nature* 410:655–660.

Enquist, B.J., West, G.B., Charnov, E.L., and Brown, J.H. 1999. Allometric scaling of production and life-history variation in vascular plants. *Nature* 401:907–911.

Etienne, R.S., and Olff, H. 2004. How dispersal limitation shapes species-body size distributions in local communities. *American Naturalist* 163: 69–83.

Fahrig, L. 2001. How much habitat is enough? *Biological Conservation* 100: 65–74.

Fahrig, L. 2002. Effect of habitat fragmentation on the extinction threshold: a synthesis. *Ecological Applications* 12:346–353.

Finlay, B.J., and Fenchel, T. 2001. Protozoan community structure in a fractal soil environment. *Protist* 152:203–218.

Fitter, A.H., and Stickland, T.R. 1992. Fractal characterization of root system architecture. *Functional Ecology* 6:632–635.

Francis, A.P. and Currie, D.J. 2003. A globally consistent richness-climate relationship for angiosperms. *American Naturalist* 161:521–536.

Fukami, T. and Wardle, D.A. 2005. Long-term ecological dynamics: reciprocal insights from natural and anthropogenic gradients. *Proceedings of the Royal Society, Series B* 272: 2105–2115.

Gaston, K.J., Chown, S.L., and Mercer, R.D. 2001. The animal species-body size distribution of Marion Island. *Proceedings of the National Academy of Sciences* 98:14493–14496.

Gause, G.F. 1934. *The struggle for existence.* Hafner, New York.

Gavrilets, S. 2004. *Fitness landscapes and the origin of species.* Princeton University Press, Princeton, New Jersey.

Giacomelli, F., Wiener, J., Kruskal, J.B., Pomeranz, J.V., and Loud, A.V. 1971. Subpopulations of blood lymphocytes demonstrated by quantitative cytochemistry. *Journal of Histochemical Cytochemistry* 19:426–433.

Gillooly, J.F., Allen, A.P., West, G.B., and Brown, J.H. 2005. The rate of DNA evolution: effects of body size and temperature on the molecular clock. *Proceedings of the National Academy of Sciences* 102:140–145.

Gotelli, N.J., and Rohde, K. 2002. Co-occurrence of ectoparasites of marine fishes: A null model analysis. *Ecology Letters* 5:86–94.

Grace, J.B. 1999. The factors controlling species density in herbaceous plant communities: an assessment. *Perspectives in Plant Ecology, Evolution and Systematics* 2:1–28.

Grant, P.R. 1986. *Ecology and evolution of Darwin's finches.* Princeton University Press, Princeton, New Jersey.

Green, J.L., Harte, J., and Ostling, A. 2003. Species richness, endemism, and abundance patterns: Tests of two fractal models in a serpentine grassland. *Ecology Letters* 6:919–928.

Greig-Smith P. 1983. *Quantitative plant ecology.* 3rd ed. Blackwell, Oxford, UK.

Grime, J.P. 1979. *Plant strategies and vegetation processes.* Wiley, Chichester, UK.

Gross, J.E., Shipley, L.A., Hobbs, N.T., Spalinger, D.E., and Wunder, B.A. 1993. Functional response of herbivores in food-concentrated patches: tests of a mechanistic model. *Ecology* 74:778–791.

Gross, K.L., Willig, M.R., Gough, L., Inouye, R., and Cox, S.B. 2000. Patterns of species density and productivity at different spatial scales in herbaceous plant communities. *Oikos* 89:417–427.

Grover, J.P. 1990. Resource competition in a variable environment: phytoplankton growing according to Monod's model. *American Naturalist* 136:771–789.

Gunnarsson, B. 1992. Fractal dimension of plants and body size distribution in spiders. *Functional Ecology* 6:636–641.

Halley, J.M., Hartley, S., Kallimanis, A.S., Kunin, W.E., Lennon, J.J., Sgardelis, S.P. 2004. Uses and abuses of fractal methodology in ecology. *Ecology Letters* 7:254–271.

Hanski, I., and Cambefort, Y., eds. 1991. *Dung beetle ecology.* Princeton University, Princeton, New Jersey.

Hardin, G. 1960. The competitive exclusion principle. *Science* 131:1292–1297.

Harrison, S., Davies, K.F., Safford, H.D., and Viers, J.H. 2006. Beta diversity and the scale-dependence of the productivity-diversity relationship: A test in the Californian serpentine flora. *Journal of Ecology* 94:110–117.

Harte, J., Conlisk, E., Ostling, A., Green, J.L., and Smith, A.B. 2005. A theory of spatial structure in ecological communities at multiple spatial scales. *Ecological Monographs* 75:179–197.

Harte, J., Zillio, T., Conlisk, E., and Smith, A.B. 2008. Maximum entropy and the state variable approach to macroecology. *Ecology* 89:2700–2711.

Hartley, W.J., and Kunin, W.E. 2003. Scale-dependence of rarity, extinction risk and conservation priority. *Conservation Biology* 17:1559–1570.

Hartley, S., Kunin, W.E., Lennon, J.J., and Pocock, M.J.O. 2005. Coherence and discontinuity in the scaling of species' distribution patterns. *Proceedings of the Royal Society of London, Series B* 271:81–88.

Haskell, J.H., Ritchie M.E., and Olff, H. 2002. Fractal geometry predicts varying body size scaling relationships for mammal and bird home ranges. *Nature* 418:527–530.

Hastings, H.M. and Sugihara, G.1993. *Fractals: a user's guide for the natural sciences.* Oxford University, Oxford, UK.

Havel, J.E., and Shurin, J.B. 2004. Mechanisms, effects, and scales of dispersal in freshwater zooplankton. *Limnology and Oceanography* 49:1229–1238.

Hawkins, B.A., Field, R., Cornell, H.V., Currie, D.J., Guegan, J.F., Kaufman, D.M., Kerr, J.T., Mittelbach, G.G., Oberdorff, T., O'Brien, E.M., Porter, E.E., and Turner, J.R.G. 2003. Energy, water, and broad-scale geographic patterns of species richness. *Ecology* 84:3105–3117.

Hawkins, C.P., and MacMahon J.A. 1989. Guilds: the multiple meanings of a concept. *Annual Review of Entomology* 34:423–451.

Hoddle, M.S. 2003. The effect of prey species and environmental complexity on the functional response of *Franklinothrips orizabensis*: A test of the fractal foraging model. *Ecological Entomology* 28: 309–318.

Hoeck, H.N. 1989. Demography and competition in hyrax: a 17-year study. *Oecologia* 79:353–360.

Holland, J.D., Fahrig, L., and Cappucino, N. 2005. Body size affects the spatial scale of beetle-habitat interactions. *Oikos* 110:101–108.

Holling, C.S. 1959. Some characteristics of simple types of predation and parasitism. *Canadian Entomologist* 91: 385–398.

Holt R.D., and Lawton J.H. 1993. Apparent competition and enemy-free space in insect host-parasitoid communities. *American Naturalist* 142: 623–645.

Holt R.D., Grover J., and Tilman D. 1994. Simple rules for interspecific dominance in systems with exploitative and apparent competition. *American Naturalist* 144:741–771.

Holt, R.D. 1987. Prey communities in patchy environments. *Oikos* 50:276–289.

Horn, H.S. 1975. Markovian properties of forest succession. Pages 196–211 in Cody, M. L., and Diamond, J.M., eds. *Ecology and evolution of communities*. Harvard University,Cambridge, Massachusetts.

Hoyle, M., and Harborne, A. R. 2005. Mixed effects of habitat fragmentation on species richness and community structure in a microarthropod microecosystem. *Ecological Entomology* 30:684–691.

Hsu, S.B., Hubbell, S.P., and Waltman, P. 1977. A mathematical theory for single-nutrient competition in continuous cultures of microorganisms. *SIAM Journal of Applied Mathematics* 32:366–383.

Hubbell, S.J. 2001. *The unified neutral theory of biodiversity and biogeography*. Princeton University, Princeton, New Jersey.

Hubbell S.J. 2006. Neutral theory and the evolution of ecological equivalence. *Ecology* 87:1387–1398.

Huberty, L.E., Gross, K.L., and Miller, C.J. 1998. Effects of nitrogen addition on successional dynamics and species diversity in Michigan old-fields. *Journal of Ecology* 86:794–803.

Huisman, J., and Weissing, F.J. 1994. Light-limited growth and competition for light in well-mixed aquatic environments: an elementary model. *Ecology* 75:507–520.

Huisman, J., and Weissing, F.J. 1999. Biodiversity of plankton by species oscillations and chaos. *Nature* 402:407–410.

Huisman, J., and Weissing, F.J. 2001. Fundamental unpredictability in multispecies competition. *American Naturalist* 157:488–494.

Hulme, P.E. 1994. Seedling herbivory in grassland: Relative impact of vertebrate and invertebrate herbivores. *Journal of Ecology* 82:873–880.

Hulme, P.E. 1996. Herbivores and the performance of grassland plants: A comparison of arthropod, mollusc and rodent herbivory. *Journal of Ecology* 84:43–51.

Huston, M.A. 1994. *Biological diversity: the coexistence of species on changing landscapes.* Cambridge University Press, Cambridge, UK.

Hutchinson, G.E. 1957. Concluding remarks. *Cold Spring Harbor Symposia on Quantitative Biology* 22:414–427.

Hutchinson, G.E. 1959. Homage to Santa Rosalia or Why are there so many kinds of animals? *American Naturalist* 93:145–159.

Hutchinson, G.E. 1965. *The ecological theater and the evolutionary play.* Yale University Press, New Haven, Connecticut.

Hutchinson, G.E., and MacArthur R.H. 1959. A theroetical ecological model of size distributions among species of animals. *American Naturalist* 93:117–125.

Huxley, J. 1942. *Evolution: the modern synthesis.* Harper, New York.

Johnson, G.D., Tempelman, A., and Patil, G.P. 1996. Fractal-based methods in ecology: a review for analysis at multiple spatial scales. *Coenoses* 10:123–131.

Kaiser, H. 1983. Small scale spatial heterogeneity influences predation success in an unexpected way: model experiments on the functional response of predatory mites (Acarina). *Oecologia* 56:249–256.

Karlson, R.H., and Cornell, H.V. 2002. Species richness of coral assemblages: Detecting regional influences at local spatial scales. *Ecology* 83: 452–463.

Karst, J., Gilbert, B., and Lechowicz, M.J. 2005. Fern community assembly: the role of chance and environment at local and intermediate scales. *Ecology* 86:2473–2486.

King, A.W., and With, K.A. 2002. Dispersal success on spatially structured landscapes: when do spatial pattern and dispersal behavior really matter? *Ecological Modelling* 147:23–39.

Kingdon, J. 1984. *East African mammals: an atlas of evolution in Africa.* University of Chicago Press, Chicago.

Kingsland, S. 1988. *Modeling nature: episodes in the history of population ecology.* University of Chicago Press, Chicago.

Kinzig, A.P., Levin, S.A., Dushoff, J., and Pacala, S. 1999. Limiting similarity, species packing, and system stability for hierarchical competition-colonization models. *American Naturalist* 153:371–383.

Klausmeier, C., and Tilman, D. 2002. Spatial models of competition. Pages 43–78 in Sommer, U., and Worm, B., eds. *Competition and coexistence.* Springer, Berlin.

Korcak, J. 1938. Deux types fondamentaux de distribution statistique. *Bulletin de l'Institute International de Statistique* III: 295-299.

Kunin, W.E. 1997. Sample shape, spatial scale and species counts: Implications for reserve design. *Biological Conservation* 82:369–377.

Kunin, W.E. 1998. Extrapolating species abundance across spatial scales. *Science* 281:1513–1515.

Kunin, W.E., and Lennon, J.J. 2004. Spatial scale and species diversity-building species-area curves from species incidence. Pages 89–108 in Shachak, M., Gosz, J.R., Pickett, S.T., Perevolotsky, A., eds. *Biodiversity in drylands: toward a unified framework*. Oxford, UK.

Lafferty, K.D., Sammond, D.T., and Kuris, A.M. 1994. Analysis of larval trematode communities. *Ecology* 75:2275–2285.

Lambers, H., Chapin, F.S. III, Pons, T.L. 1998. *Plant physiological ecology*. Springer NY.

Lawton, J.H., and Strong, D.R. 1981. Community patterns and competition on folivorous insects. *American Naturalist* 118:317–338.

Lawton, J.H. 1999. Are there general laws in ecology? *Oikos* 84:177–192.

Leibold, M.A. 1989. Resource edibility and the effects of predators and productivity on the outcome of trophic interactions. *American Naturalist* 134:922–949.

Leibold, M.A. 1995. The niche concept revisited: Mechanistic models and community context. *Ecology* 76:1371–1382.

Leibold, M.A. 1996. A graphical model of keystone predators in food webs: trophic regulation of abundance, incidence, and diversity patterns in communities. *American Naturalist* 147:784–812.

Leibold, M.A. 1998. Similarity and local co-existence of species in regional biotas. *Evolutionary Ecology* 12:95–110.

Leibold, M.A., Holyoak, M., Mouquet, N., Amarasekare, P., Chase, J.M., Hoopes, M.F., Holt, R.D., Shurin, J.B., Law, R., Tilman, D., Loreau, M., and Gonzalez, A. 2004. The metacommunity concept: a framework for multi-scale community ecology. *Ecology Letters* 7:601–613.

Leibold, M.A., and McPeek, M.A. 2006. Coexistence of the niche and neutral perspectives in community ecology. *Ecology* 87:1399–1410.

Levin, S.A. 1992. The problem of pattern and scale in ecology. *Ecology* 73:1943–1967.

Levins, R. 1962. Theory of fitness in a heterogeneous environment. I. The fitness set and adaptive function. *American Naturalist* 96:361–373.

Levins, R. 1968. *Evolution in changing environments*. Princeton University Press, Princeton, New Jersey.

Li, B. 2000. Fractal geometry applications in description and analysis of patch patterns and patch dynamics. *Ecological Modeling* 132:33–50.

Linder, H.P., and Hardy, C.R. 2004. Evolution of the species-rich Cape flora. *Philosophical Transactions: Biological Sciences* 359:1623–1632.

Lotka, A.J. 1925. *Elements of physical biology*. Williams and Wilkins, Baltimore Maryland.

Lovejoy, S., Currie, W.J.S., Tessier, Y., Claereboudt, M.R., Bourget, E., Roff, J.C., and Schertzer, D. 2001. Universal multifractals and ocean patchiness: phytoplankton, physical fields and coastal heterogeneity. *Journal of Plankton Research* 23:117–141.

MacArthur, R.H. 1958. Population ecology of some warblers of northeastern coniferous forests. Ecology 39:599–619.

MacArthur, R.H. 1965. Patterns of species diversity. *Biology Reviews* 40:510–533.

MacArthur, R.H. 1969. Species packing and what interspecies competition minimizes. *Proceedings of the National Academy of Sciences* 64:1369–1371.

MacArthur, R.H. 1970. Species packing and competitive equilibrium for many species. *Theoretical Population Biology* 1:1–11.

MacArthur, R. H. 1972. *Geographical ecology: patterns in the distribution of species.* Princeton University Press, New Jersey.

MacArthur, R.H., and Levins, R. 1964. Competition, habitat selection, and character displacement in a patchy environment. *Proceedings of the National Academy of Sciences* 51:1207–1210.

MacArthur R.H., and Levins, R. 1967. The limiting similarity, convergence, and divergence of coexisting species. *American Naturalist* 101:377–385.

MacArthur, R.H., and Pianka, E.R. 1966. On the optimal use of a patchy environment. *American Naturalist* 100:603–609.

MacArthur, R.H., and Wilson, E.O. 1967. *The theory of island biogeography.* Princeton University Press, Princeton, New Jersey.

Magurran, A. 2004. *Measuring biological diversity,* Blackwell, London, UK.

Magurran, A., and Henderson, P.A. 2003. Explaining the excess of rare species in species abundance distributions. *Nature* 422:714–716.

Mahlaba, T.A., and Perrin, M.R. 2003. Population dynamics of small mammals at Mlawula, Swaziland. *African Journal of Ecology* 41:317–323.

Mandelbrot, B.B. 1982. *The fractal geometry of nature.* Freeman, New York.

Marquet, P.A., and Taper, M.L. 1998. On size and area: patterns of mammalian body size extremes across landmasses. *Evolutionary Ecology* 12:127–139.

Maurer, B.A. 1998. The evolution of body size in birds. II. The role of reproductive power. *Evolutionary Ecology* 12:935–944.

Maurer, B.A., and Brown, J.H. 1988. Distribution of biomass and energy use among species of North American terrestrial birds. *Ecology* 69:1923–1932.

May, R.M. 1974. *Stability and complexity in model ecosystems.* Princeton University Press, Princeton, New Jersey.

May, R.M. 1975. Patterns of species abundance and diversity. Pages 81–120 in Cody, M.L., and Diamond, J.M., eds. *Ecology and evolution of communities.* Harvard University Press, Cambridge, Massachusetts.

May, R.M. 1976. *Theoretical ecology: principles and applications.* Sinauer Press, Sunderland, Massachusetts.

May, R.M. 1988. How many species are there on Earth? *Science* 241:1441–1449.

McAbendroth, L., Ramsay, P.M., Foggo, A., Rundle, S.D., Bilton, D.T. 2005. Does macrophyte fractal complexity drive invertebrate diversity, biomass and body size distributions? *Oikos* 111:279–290.

McCarthy, M.A., Thompson, C.J., and Williams, N.S.G. 2006. Logic for designing nature reserves for multiple species. *American Naturalist* 167:717–727.

McClatchie, S., Greene, C.H., Macaulay, M.C., and Sturley, D.R.M. 1994. Spatial and temporal variability of Antarctic krill: Implications for stock assessment. *I.C.E.S. Journal of Marine Science* 51:11–18.

McClure, M.F., Bissonette, J.A., and Conover, M.R. 2005. Migratory strategies, fawn recruitment, and winter habitat use by urban and rural mule deer (*Odocoileus hemionus*). *European Journal of Wildlife Research* 51:170–177.

McGarigal, K., McComb, W.C. 1995. Relationships between landscape structure and breeding birds in the Oregon Coast Range. *Ecological Monographs* 65, 235–260.

McGill, B.J. 2003. A test of the unified neutral theory of biodiversity. *Nature* 422:881–885.

McGill B.J., Maurer, B.A., and Weiser, M.D. 2006. Empirical evaluation of neutral theory. *Ecology* 87:1411–1423.

Mehner, T., and Thiel, R. 1999. A review of predation impact by fish on zooplankton in fresh and brackish waters of the temperate Northern Hemisphere. *Environmental Biology of Fishes* 56:169–181.

Milne, B.T. 1992. Spatial aggregation and neutral models in fractal landscapes. *American Naturalist* 139:32–57.

Milne, B.T. 1997. Applications of fractal geometry in wildlife biology. Pages 32–69 in Bissonette, J.A., ed. *Wildlife and landscape ecology*. Springer, New York.

Milne, B.T., Turner M.G., Wiens, J.A., and Johnson, A.R. 1992. Interactions between the fractal geometry of landscapes and allometric herbivory. *Theoretical Population Biology* 41:337–353.

Mittelbach, G.G., Steiner, C.F., Scheiner, S.M., Gross, K.L., Reynolds, H.L., Waide, R.B., Willig, M.R., Dodson, S.I., and Gough, L. 2001. What is the observed relationship between species richness and productivity? *Ecology* 82:2381–2396.

Morse, D.R., Lawton, J.H., Dodson, M.M., and Williamson, M.H. 1985. Fractal dimension of vegetation and the distribution of arthropod body lengths. *Nature* 314:731–733.

Mouquet, N., and Loreau, M. 2003. Community patterns in source-sink metacommunities. *American Naturalist* 162:544–557.

Murdoch, W.W., Briggs, C.J., and Nisbet, R.M. 2003. *Consumer-resource dynamics*. Princeton University Press, Princeton, New Jersey.

Naeem, S. 1990. Resource heterogeneity and community structure: a case study in *Heliconia imbricata*, Phytotelmata. *Oecologia* 84:29–38.

Olff, H., and Ritchie, M.E. 1998. Herbivore effects on grassland plant diversity. *Trends in Ecology and Evolution* 13:261–265.

Olff, H., and Ritchie, M.E. 2002. Fragmented nature: consequences for biodiversity. *Landscape and Urban Planning* 58:83–92.

Olff, H., Ritchie, M.E., and Prins, H.H.T. 2002. Global determinants of diversity in large herbivores. *Nature* 415:901–904.

O'Neill, R.V., DeAngelis, D.L., Waide, J.B., and Allen, T.F.H. 1986. *A hierarchical concept of ecosystems.* Princeton University Press, Princeton, New Jersey.

Ostling, A., Harte, J., Green, J.L., and Kinzig, A.P. 2003. A community-level fractal property produces power-law species-area relationships. *Oikos* 103:218–224.

Otte, D., and Endler, J.A., eds. 1989. *Speciation and its consequences.* Sinauer, Sunderland, Massachusetts.

Pacala, S.W. 1982. The evolution of resource partitioning in a multidimensional resource space. *Theoretical Population Biology* 22:127–145.

Pacala, S.W., and Tilman D. 1994. Limiting similarity in mechanistic and spatial models of plant competition in heterogeneous environments. *American Naturalist* 143:222–257.

Paine, R.T. 1966. Food web complexity and species diversity. *American Naturalist* 100:65–75.

Paine, R.T. 1992. Food-web analysis through field measurement of per capita interaction strength. *Nature* 355:73–75.

Palmer, M.W. 1992. The coexistence of species in fractal landscapes. *American Naturalist* 139:375–397.

Park, T. 1948. Experimental studies of interspecies competition. I. Competition between populations of the flour beetles, *Tribolium confusum* Duvall and *Tribolium castaneum* Herbst. *Ecological Monographs* 18:267–307.

Peters, R.H. 1983. *The ecological implications of body size.* Cambridge University Press, Cambridge, UK.

Phillips, M.L., Clark, W.R., Nusser, S.M., Sovada, M.A., and Greenwood, R.J. 2004. Analysis of predator movement in prairie landscapes with contrasting grassland composition. *Journal of Mammalogy* 85:187–195.

Pitt, W.C., and Ritchie, M.E. 2002. Influence of prey distribution on the functional response of lizards. *Oikos* 96:157–163.

Preston, F.W. 1962. The canonical distribution of commonness and rarity. *Ecology* 43:185–215.

Pulliam, H.R. 1974. On the theory of optimal diets. *American Naturalist* 108:59–75.

Radford, J.Q., Bennett, A.P., and Cheers, G.C. 2005. Landscape-level thresholds of habitat cover for woodland-dependent birds. *Biological Conservation* 124:317–337.

Rajaniemi, T.K. 2003. Explaining productivity-diversity relationships in plants. *Oikos* 101:449–457.

Rasiah, V., and Aylmore, L.A.G. 1998. Estimating microscale spatial distribution of conductivity and pore continuity using computed tomography. *Soil Science Society of America Journal* 62:1197–1202.

Reich, P.B., Hobbie, S.E., Lee, T., Ellsworth, D.S., West, J.B., and Tilman, D. et al. 2006. Nitrogen limitation constrains sustainability of ecosystem response to CO_2. *Nature* 440:922–925.

Reich, P.B., Knops, J., Tilman, D., Craine, J., Ellsworth, D., and Tjoelker, M. et al. 2001. Plant diversity enhances ecosystem responses to elevated CO_2 and nitrogen deposition. *Nature* 410:809–812.

Ricklefs, R.E., and Schluter, D., eds. 1994. *Species diversity in ecological communities.* University of Chicago Press, Chicago.

Ritchie, M.E. 1996. Interaction of temperature and resources in population dynamics: an experimental test of theory. Pages 79–92 in Floyd, R.B., Sheppard, A.W., and DeBarro, P.J., eds. *Frontiers in Population Ecology.* CSIRO Press, Melbourne.

Ritchie, M.E. 1997. Population dynamics in a landscape context: sources, sinks, and metapopulations. Pages 160–184 in Bissonette, J.A. ed. *Wildlife and Landscape Ecology*, Springer, New York.

Ritchie, M.E. 1998. Scale-dependent foraging and patch choice in fractal environments. *Evolutionary Ecology* 12:309–330.

Ritchie, M.E. 2000. Nitrogen limitation and trophic vs. abiotic influences on insect herbivores in a temperate grassland. *Ecology* 81:1601–1612.

Ritchie, M.E. 2002. Competition and coexistence in mobile animals. Pages 112–135 in Sommer, U., and Worm, B., eds. *Competition and coexistence.* Springer, Berlin.

Ritchie, M.E., and Olff, H. 1999. Spatial scaling laws yield a synthetic theory of biodiversity. *Nature* 400:557–560.

Ritchie, M.E. and H. Olff. 2005. The scaling of resource partitioning and biodiversity in dryland environments. Pages 206–219 in Schachak, M., and Waide, R.L., eds. *Biodiversity in dryland environments.* Oxford University Press.

Ritchie, M.E., and Tilman, D. 1992. Interspecific competition among grasshoppers and their effect on plant abundance in experimental field environments. *Oecologia* 89:524–532.

Ritchie, M.E., and Tilman, D. 1993. Predictions of species interactions from consumer-resource theory: experimental tests with plants and grasshoppers. *Oecologia* 94: 516–527.

Ritchie, M.E., Wolfe, M.L., and Danvir, R. 1994. Predation on artificial sage grouse nests in treated and untreated sagebrush. *Great Basin Naturalist*, 54:122–129.

Robinson, W.D., Brawn, J.D., and Robinson, S.K. 2000. Forest bird community structure in central Panama: influence of spatial scale and biogeography. *Ecological Monographs* 70:209–235.

Rohde, K. 2001. Spatial scaling laws may not apply to most animal species. *Oikos* 99:499–504.

Root, R.B. 1996. Herbivore pressure on goldenrods (*Solidago altissima*): Its variation and cumulative effects. *Ecology* 77:1074–1087.

Rosenzweig, M.L. 1966. Community structure in sympatric carnivora. *Journal of Mammology* 47:602–612.

Rosenzweig, M.L. 1995. *Species diversity in space and time.* Cambridge University Press, Cambridge, UK.

Rothhaupt, K.O. 1988. Mechanistic resource competition theory applied to laboratory experiments with zooplankton. *Nature* 333:660–662.

Roughgarden, J. 1974. Species packing and the competition function with illustrations from coral reef fish. *Theoretical Population Biology* 5:163–186.

Safford, H.D., Rejmanek, M., and Hadac, E. 2001. Species pools and the "hump-back" model of plant species diversity: An empirical analysis at a relevant spatial scale. *Oikos* 95:282–290.

Sala, O.E., Chapin, F.S., III, Armesto, J.J., Berlow, E., Bloomfield, J., Dirzo, R., Huber-Sanwald, E., Huenneke, L.F., Jackson, R.B., Kinzig, A., Leemans, R., Lodge, D.M., Mooney, H.A., Oesterheld, M., Poff, N.L., Sykes, M.T., Walker, B.H., Walker, M., and Wall, D.H. 2000. Global biodiversity scenarios for the year 2100. *Science* 287:1770–1774.

Sarr, D.A., Hibbs, D.E., and Huston, M.A. 2005. A hierarchical perspective of plant diversity. *Quarterly Review of Biology* 80:187–212.

Schmidt-Nielsen, K. 1983. *Scaling: why is animal size so important?* Cambridge University, U.K.

Schmitz, O.J. 1992. Exploitation in model food chains with mechanistic consumer-resource dynamics. *Theoretical Population Biology* 41:161–183.

Schmitz, O.J. 2003. Top predator control of plant biodiversity and productivity in an old-field ecosystem. *Ecology Letters* 6:156–163.

Schmitz, O.J., Beckerman, A.P., and Litman, S. 1997. Functional responses of adaptive consumers and community stability with emphasis on the dynamics of plant-herbivore systems. *Evolutionary Ecology* 11:773–784.

Schmitz, O.J., and Suttle, K.B. 2001. Effects of top predator species on direct and indirect interactions in a food web. *Ecology* 82:2072–2081.

Schoener, T.W. 1965. The evolution of bill size differences among sympatric species of birds. *Evolution* 19:189–213.

Schoener, T.W. 1971. Theory of feeding strategies. *Annual Review of Ecology and Systematics* 2:369–404.

Schoener, T.W. 1973. Population growth regulated by intraspecific competition for energy or time: some simple representations. *Theoretical Population Biology* 4:56–84.

Schoener, T.W. 1974. Resource partitioning in ecological communities. *Science* 185:27–39.

Schoener, T.W. 1976. Alternatives to Lotka-Volterra competition: models of intermediate complexity. *Theoretical Population Biology* 10:309–333.

Schoener, T.W. 1978. Effects of density-restricted food encounter on some single-level competition models. *Theoretical Population Biology* 13: 365–381.

Schoener, T.W. 1983. Field experiments on interspecific competition. *American Naturalist* 122:240–285.

Schoener, T.W. 1984. Sized differences among sympatric, bird-eating hawks: a worldwide survey. Pages 254–281 in Strong, D.R., Simberloff, D., Able, L.G., and Thistle, A.B. eds. *Ecological communities: conceptual issues and the evidence.* Princeton University Press, Princeton, New Jersey.

Schooley, R.L., and Wiens, J.A. 2001. Dispersion of kangaroo rat mounds at multiple scales in New Mexico, USA. *Landscape Ecology* 16:267–277.

Seuront, L., and Lagadeuc, Y. 2001. Multiscale patchiness of the calanoid copepod temora longicornis in a turbulent coastal sea. *Journal of Plankton Research* 23:1137–1145.

Shipley, B., Vile, D., and Garnier, E. 2007. From plant traits to plant communities: a statistical mechanistic approach to biodiversity. *Science* 316: 201–203.

Shorrocks, B., Marsters, J., Ward, I., and Evennett, P.J. 1991.The fractal dimension of lichens and the distribution of arthropod body lengths. *Functional Ecology* 5:457–460.

Shurin, J.B., and Allen, E.G. 2001. Effects of competition, predation, and dispersal on species richness at local and regional scales. *American Naturalist* 158:624–637.

Siemann, E., Tilman, D., and Haarstad, J. 1996. Insect species diversity, abundance and body size relationships. *Nature* 380:704–706.

Sih, A., Crowley, P., McPeek, M., Petranka, J., and Strohmeier, K. 1985. Predation, competition, and prey communities: A review of field experiments. *Annual Review of Ecology and Systematics* 16:269–311.

Simberloff, D. 1983. Competition theory, hypothesis-testing, and other community ecological buzzwords. *American Naturalist* 122:626–635.

Simberloff, D. 2004. Community ecology: is it time to move on? *American Naturalist* 163:787–799.

Simberloff, D., and Boecklen, W. 1981. Santa Rosalia reconsidered: Size ratios and competition. *Evolution* 35:1206–1228.

Simberloff, D., and Wilson, E.O. 1969. Experimental zoogeography of islands. The colonization of empty islands. *Ecology* 50:278–296.

Sims, D.W., Southall, E.J., Humphries, N.E., Hays, G.C., Bradshaw, C.J.A., et al. 2008. Scaling laws of marine predator search behaviour. *Nature* 451:1098–1102.

Sinclair, A.R.E., Mduma, S., and Brashares, J.S. 2003. Patterns of predation in a diverse predator-prey system. *Nature* 424:288–290.

Slatkin, M. 1980. Ecological character displacement. *Ecology* 61:163–177.

Smith, V.H., and Bennett, S.J. 1999. Nitrogen: phosphorus supply ratios and phytoplankton community structure in lakes. *Archiv für Hydrobiologie* 146:37–53.

Snyder, R.E., and Chesson, P. 2004. How the spatial scales of dispersal, competition, and environmental heterogeneity interact to affect coexistence. *American Naturalist* 164:633–650.

Sogard, S.M. 1997. Size-selective mortality in the juvenile stage of teleost fishes: a review. *Bulletin of Marine Science* 60:1129–1157.

Sommer, U. 2002. Competition and coexistence in plankton communities. Pages 79–105 in Sommer, U. and Worm, B., eds. 2002. *Competition and coexistence*. Springer, New York.

Sommer, U., and Worm, B., eds., 2002. *Competition and coexistence*. Springer, New York.

Stephens, D.W. and Krebs, J.R. 1986. *Foraging theory*. Princeton University Press, Princeton, New Jersey.

Storch, D., Šizling, A., Reif, J., Polechová, J., Šizlingová E., and Gaston, K.J. 2008. The quest for a null model for macroecological patterns: geometry of species distributions at multiple spatial scales. *Ecology Letters* 11:771–784.

Strobeck, C. 1972. N species competition. *Ecology* 54:650–654.

Terborgh, J., Robinson, S.K., Parker, T.A., Munn, C.A., and Pierpont, N. 1990. Structure and organization of an Amazonian forest bird community. *Ecological Monographs* 60:213–238.

Thompson, J. 1994. *The coevolutionary process*. University of Chicago Press, Chicago.

Tilman, D. 1976. Ecological competition between algae: experimental confirmation of resource-based competition theory. *Science* 192:463–465.

Tilman, D. 1982. *Resource competition and community structure*. Princeton University Press, Princeton, New Jersey.

Tilman, D. 1988. *Plant strategies and the dynamics and structure of communities*. Princeton University Press, Princeton, New Jersey.

Tilman, D. 1990. Constraints and tradeoffs: toward a predictive theory of competition and succession. *Oikos* 58:3–15.

Tilman, D. 1994. Competition and biodiversity in spatially structured habitats. *Ecology* 75:2–16.

Tilman, D., and Wedin, D. 1991. Dynamics of nitrogen competition between successional grasses. *Ecology* 72:1038–1049.

Trzcinski, M.K., Fahrig, L., and Merriam, G. 1999. Independent effects of forest cover and fragmentation on the distribution of forest breeding birds. *Ecological Applications* 9:586–593.

Tsuda, A. 1995. Fractal distribution of an oceanic copepod *Neocalanus cristatus* in the subarctic Pacific. *Journal of Oceanography*. 51:261–266.

Turchin, P. 1996. Fractal analyses of animal movement: A critique. *Ecology* 77:2086–2090.

Turcotte, R.H. 1995. Scaling in geology: landforms and earthquakes. *Proceedings of the National Academy of Sciences* 95:6697–6704.

Van Soest, P.J., ed. 1994. *Nutritional ecology of the ruminant*. 2nd ed. Cornell University, Ithaca, New York.

Vandermeer, J.H. 1972. Niche theory. *Annual Review of Ecology and Systematics* 3:107–132.

Vandermeer, J.H., Perfecto, I., and Philpott, S.M. 2008. Clusters of ant colonies and robust criticality in a tropical agroecosystem. *Nature* 451: 457–459.

Vance, M.D., Fahrig, L., and Flather, C.H. 2003. Effect of reproductive rate on minimum habitat requirements of forest-breeding birds. *Ecology* 84: 2643–2653.

Venterink, H.O., Wassen, M.J., Verkroost, A.W.M., and de Ruiter, P.C. 2003. Species richness-productivity patterns differ between N-, P-, and K-limited wetlands. *Ecology* 84:2191–2199.

Vickery, P.D., Hunter, M.L., and Melvin, S. 1994. Effects of habitat area on the distribution of grassland birds in Maine. *Conservation Biology* 8: 1087–1097.

Volkov, I., Banavar, J.R., Hubbell, S.P., and Maritan, A. 2003. Neutral theory and relative species abundance in ecology. *Nature* 424:1035–1037.

Volterra, V. 1926. Fluctuations in the abundance of a species considered mathematically. *Nature* 118:558–560.

Voss, R.F. 1986. The measurement and characterization of random fractals. *Physica Scripta* T13:27–32.

Waide, R.B., Willig, M.R., Steiner, C.F., Mittelbach, G., Gough, L., Dodson, S.I., Juday, G.P., and Parmenter, R. 1999. The relationship between productivity and species richness. *Annual Review of Ecology and Systematics* 30:257–300.

Ward, D. 1992. The role of satisficing in foraging theory. *Oikos* 63:312–317.

Warren, J.D., Santora, J., and Demer, D. 2005. Where should we eat? Predator and prey spatial relationships in the Antarctic krill ecosystem. *Journal of the Acoustical Society of America* 118:1910.

Waters, R.L., and Mitchell, J.G. 2002. Centimetre-scale spatial structure of estuarine in vivo fluorescence profiles. *Marine Ecology Progress Series* 237:51–63.

Waters, R.L., Mitchell, J.G., and Seymour, J. 2003. Geostatistical characterization of centimetre-scale spatial structure of in vivo fluorescence. *Marine Ecology Progress Series* 251:49–58.

Welch, R., and Kaiser, D. 2001. Cell behavior in traveling wave patterns of myxobacteria. *Proceedings of the National Academy of Sciences* 98: 14907–14912.

Werner, E.E. 1994. Ontogenetic scaling of competitive relations: size-dependent effects and responses in two anuran larvae. *Ecology* 75: 197–213.

Werner E.E., and Gilliam, J.F. 1984. The ontogenetic niche and species interactions in size-structured populations. *Annual Review of Ecology and Systematics* 15:393–425.

West, G.B., Brown J.H., and Enquist, B.J. 1997. A general model for the origin of allometric scaling laws in biology. *Science* 276:122–126.

West, G.B., Brown, J.H., and Enquist, B.J. 1999. A general model for the structure and allometry of plant vascular systems. *Nature* 400: 664–667.

Whittaker, R.H. 1975. *Communities and ecosystems.* 2nd ed. Macmillan, New York.

Wiens, J.A., Crist, T.O., and Milne, B.T. 1993. On quantifying insect movements. *Environmental Entomology* 22:709–715.

Wiens, J.A., and Milne, B.T. 1989. Scaling of 'landscapes' in landscape ecology, or, landscape ecology from a beetle's perspective. *Landscape Ecology* 3:87–96.

Wilson, E.O. 1992. *The diversity of life.* Penguin, New York.

Wilson, W.G., Lundberg, P., Vazquez, D.P., Shurin, J.B., Smith, M.D., and Langford, W., et al. 2003. Biodiversity and species interactions: Extending Lotka-Volterra community theory. *Ecology Letters* 6:944–952.

With, K.A. 1994. Ontogenetic shifts in how grasshoppers interact with landscape structure: An analysis of movement patterns. *Functional Ecology* 8:477–485.

With, K.A., and Crist, T.O. 1995. Critical thresholds in species' responses to landscape structure. *Ecology* 76:2446–2459.

With, K.A., and King, A.W. 1999. Extinction thresholds for species in fractal landscapes. *Conservation Biology* 13:314–326.

With, K.A., and King, A.W. 2004. The effect of landscape structure on community self-organization and critical biodiversity. *Ecological Modelling* 179:349–366.

Wolfram, S.R. 2002. *A new kind of science.* Wolfram Media, Champaign, Illinois.

Wootton, J.T. 1997. Estimates and tests of per capita interaction strength: diet, abundance, and impact of intertidally foraging birds. *Ecological Monographs* 67:45–64.

Worm, B., Lotze, H.K., Hillebrand, H., and Sommer, U. 2002. Consumer versus resource control of species diversity and ecosystem functioning. *Nature* 417:848–851.

Yoshiyama, K., and Klausmeier, C.A. 2008. Optimal cell size for resource uptake in fluids: a new facet of resource competition. *American Naturalist* 171:59–70.

Zou, X., Zucca, C.P., Waide, R.B., and McDowell, W.H. 1995. Long-term influence of deforestation on tree species composition and litter dynamics of a tropical rain forest in Puerto Rico. *Forest Ecology and Management* 78:147–157.

Index

MONOGRAPHS IN POPULATION BIOLOGY
EDITED BY SIMON A. LEVIN AND HENRY S. HORN

Titles available in the series (by monograph number)